Integrated Water Treatment: Softening and Ultrafiltration

The mission of the Awwa Research Foundation (AwwaRF) is to advance the science of water to improve the quality of life. Funded primarily through annual subscription payments from over 1,000 utilities, consulting firms, and manufacturers in North America and abroad, AwwaRF sponsors research on all aspects of drinking water, including supply and resources, treatment, monitoring and analysis, distribution, management, and health effects.

From its headquarters in Denver, Colorado, the AwwaRF staff directs and supports the efforts of over 500 volunteers, who are the heart of the research program. These volunteers, serving on various boards and committees, use their expertise to select and monitor research studies to benefit the entire drinking water community.

Research findings are disseminated through a number of technology transfer activities, including research reports, conferences, videotape summaries, and periodicals.

Integrated Water Treatment: Softening and Ultrafiltration

Prepared by:
Desmond F. Lawler and **JiHyang Kweon**
The University of Texas at Austin
Department of Civil Engineering
301 East Dean Keaton Street
Austin, TX 78712-1076

Sponsored by:
Awwa Research Foundation
6666 West Quincy Avenue
Denver, CO 80235

Published by the
Awwa Research Foundation
Advancing the Science of Water®
IWA Publishing

DISCLAIMER

This study was funded by the Awwa Research Foundation (AwwaRF). AwwaRF assumes no responsibility for the content of the research study reported in this publication or for the opinions or statements of fact expressed in the report. The mention of trade names for commercial products does not represent or imply the approval or endorsement of AwwaRF. This report is presented solely for informational purposes.

Published by IWA Publishing, Alliance House, 12 Caxton Street, London SW1H 0QS, UK

Telephone: +44 (0) 20 7654 5500; Fax: +44 (0) 20 7654 5555; Email: publications@iwap.co.uk; Web: **www.iwapublishing.com**

AwwaRF report number 90937F. Originally published by AwwaRF for its subscribers in 2003. IWA Publishing version published 2004.

ISBN 1 84339 847 8

CONTENTS

TABLES

FIGURES

FOREWORD

The Awwa Research Foundation is a nonprofit corporation that is dedicated to the implementation of a research effort to help utilities respond to regulatory requirements and traditional high-priority concerns of the industry. The research agenda is developed through a process of consultation with subscribers and drinking water professionals. Under the umbrella of a Strategic Research Plan, the Research Advisory Council prioritizes the suggested projects based upon current and future needs, applicability, and past work; the recommendations are forwarded to the Board of Trustees for final selection. The foundation also sponsors research projects through the unsolicited proposal process; the Collaborative Research, Research Applications, and Tailored Collaboration programs; and various joint research efforts with organizations such as the U.S. Environmental Protection Agency, the U.S. Bureau of Reclamation, and the Association of California Water Agencies.

This publication is a result of one of these sponsored studies, and it is hoped that its findings will be applied in communities throughout the world. The following report serves not only as a means of communicating the results of the water industry's centralized research program but also as a tool to enlist the further support of the nonmember utilities and individuals. Projects are managed closely from their inception to the final report by the foundation's staff and large cadre of volunteers who willingly contribute their time and expertise. The foundation serves a planning and management function and awards contracts to other institutions such as water utilities, universities, and engineering firms. The funding for this research effort comes primarily from the Subscription Program, through which water utilities subscribe to the research program and make an annual payment proportionate to the volume of water they deliver and consultants and manufacturers subscribe based on their annual billings. The program offers a cost-effective and fair method for funding research in the public interest.

A broad spectrum of water supply issues is addressed by the foundation's research agenda: resources, treatment and operations, distribution and storage, water quality and analysis, toxicology, economics, and management. The ultimate purpose of the coordinated effort is to assist water suppliers to provide the highest possible quality of water economically and reliably. The true benefits are realized when the results are implemented at the utility level. The foundation's trustees are pleased to offer this publication as a contribution toward that end.

Edmund G. Archuleta
Chair, Board of Trustees
Awwa Research Foundation

James F. Manwaring, P.E.
Executive Director
Awwa Research Foundation

ACKNOWLEDGMENTS

Research efforts are rarely done without the assistance of many people, and this one is no exception. The authors are grateful for all of the assistance noted below.

Two utilities, the City of Austin Water and Wastewater Utility and St. Louis County Water Company, not only provided the natural waters that were used in this research, but were gracious in sharing data and operational practices of their softening plants with us.

The Project Advisory Committee was composed of Mark R. Wiesner (Rice University, Houston, TX), Scott D. Freeman (Black and Veatch, Kansas City, MO), and Craig Anderson (Florida Water Services, Orlando, FL). Mark Wiesner was particularly valuable in refining the research design, and Scott Freeman provided an excellent detailed review of the final report. The Project Officer, Stephanie Morales, was instrumental in keeping the project on track at all times.

Other faculty members at the University of Texas (UT) were helpful in completing this project. Within the Environmental and Water Resources Engineering (EWRE) program in the Dept. of Civil Engineering (where the authors are based), faculty members Lynn E. Katz, Howard M. Liljestrand, and Gerald E. Speitel Jr. gave freely of their time and advice from conception to completion, as did Douglas R. Lloyd of the Dept. of Chemical Engineering. We also appreciate being allowed to use instrumentation in other laboratories at UT, and for the faculty help in data acquisition and interpretation in each case: Kitty Milliken (scanning electron microscopy), Jose Lozano (X-ray photoelectron spectroscopy), and Ti Cao (gel permeation chromatography).

Graduate students within EWRE also provided help in many ways; particularly noteworthy are the contributions of Derek Baker, Caroline Gerwe, David Kobe, and Shay Ralls. Charles Perego, technician extraordinaire, solved many mechanical problems. Finally, the work could not have been completed without the support of JiHyeon Song and Suebin Song, husband and daughter of the second author, who accepted the long days and late nights that were required to complete the laboratory work.

EXECUTIVE SUMMARY

Membrane processes, especially low-pressure membrane processes such as microfiltration (MF) and ultrafiltration (UF), are now frequently considered for application in drinking water treatment. The biggest impediment for applying membrane processes is fouling that comes from mass flux (such as particle and organic matter) to the membrane surface and its pores due to convection flow through the membrane. The flow resistance increases as a membrane becomes fouled, so the membrane has to be cleaned periodically. Therefore, central issues for the further development and use of membrane processes are to understand the fouling mechanisms and to minimize fouling.

Pretreatment before the membrane process has become an important aspect of membrane operations, because pretreatment can reduce the concentration or change the nature of foulants in the feed flow before they deteriorate the membrane performance. Limited studies on pretreatment schemes for ultrafiltration have included coagulation followed by sedimentation or dissolved air flotation, carbon adsorption, and membrane processes with larger pore sizes. Recent studies have shown that coagulation processes substantially remove natural organic matter (NOM) and disinfection by-product (DBP) precursors in addition to particulate matter, including microorganisms. Therefore, coagulation has been investigated as a pretreatment for membrane processes because of its capability to remove the main foulants as well as the ability to use the existing infrastructure. Softening is traditionally designed to remove hardness ions from hard waters, but it also has capability to remove particles and organic matter. However, virtually no effort has been directed toward understanding the use of softening as a pretreatment before membrane processes.

RESEARCH OBJECTIVES

Many utilities throughout the central U.S. and Florida use precipitative softening, and many of those plants require modifications to meet the recent changes in regulations, which require better operation of disinfection processes. Membrane processes are one of the options considered by these softening utilities. Therefore, the objectives of the proposed research were as follows:

1. To understand the nature of the fouling mechanisms for ultrafiltration membanes when used for waters that either require softening or have been softened, and
2. To use that understanding to determine promising options for the use of precipitative softening as a pretreatment before ultrafiltration.

APPROACH

To achieve the objectives of the research, bench-scale softening jar tests and membrane filtration were involved. The source waters included two natural waters and synthetic waters that were designed to allow testing of the effects of specific constituents on both softening and membrane performance. The natural waters included those used by the City of Austin, TX (Lake Austin) and St. Louis County, MO (Missouri River). The synthetic waters included one with inorganic constituents only (simulated to match Lake Austin water), waters with simple organic components, and a water with kaolin. The synthetic inorganic water and the water with kaolin were used to investigate inorganic and particle fouling without NOM effects on fouling. The synthetic organic waters were produced with simple organic components, *i.e.*, polysaccharides, to obtain an understanding of NOM fouling.

Several fouling mechanisms could be hypothesized in softening: inorganic fouling by continuous precipitative softening, organic fouling, particulate fouling, and combined fouling by particles and organic matter. To examine the inorganic fouling, various scenarios for softening pretreatment were tested using the synthetic inorganic waters; the scenarios included lime softening alone, lime softening followed by pH adjustment, lime-soda ash softening, and lime-soda ash softening with pH adjustment. A specific softening process that achieved the least inorganic fouling on the membrane surface was then selected and used in subsequent experiments to further investigate the effects of NOM and particles on membrane performance.

Softening tests were performed with various lime doses using standard jar test conditions. Since the softening process is used as a pretreatment for ultrafiltration, the degree of softening was investigated thoroughly. The three possible levels of softening are referred to herein as standard softening, enhanced softening, and Mg softening conditions. The "standard softening" condition was chosen on the basis of softening parameters alone (i.e., the minimum lime dose to achieve excellent calcium removals). The second "enhanced softening" condition represented the maximum lime dose without forming voluminous precipitates of $Mg(OH)_2$, and the third "Mg softening" condition was a higher lime dose where the precipitation of $Mg(OH)_2$ would occur to a significant extent and remove a high percentage of the magnesium.

Ultrafiltration experiments were performed in cross-flow mode in a laboratory-scale flat sheet module. Fouling was monitored by measuring flux decline and transmembrane pressure (TMP). At the end of each experiment, some membranes were cleaned with three cleaning methods, and the clean water flux was measured after each cleaning to evaluate the extent of three fouling mechanisms: surface deposition, organic fouling, and inorganic fouling. Other fouled membranes were further investigated with scanning electron microscopy (SEM) or X-ray photoelectron spectroscopy (XPS) to understand fouling as it occurred adjacent to the membrane surface.

CONCLUSIONS

1. Inorganic fouling shows negligible effect on membrane fouling. Inorganic fouling by CaCO3 and Mg(OH)2 precipitates was investigated in three aspects: effects of the slower precipitation kinetics in softening, the degree of softening, and various scenarios of softening such as lime softening alone, lime/soda ash softening, and lime/soda ash softening with pH adjustment.
2. Ultrafiltration is not susceptible to much particulate fouling under the operational conditions in this research. Ultrafiltration showed virtually no flux decline when the synthetic inorganic water with only a clay material (without NOM) was treated. It should be noted that ultrafiltration sustained almost 100% of the clean water specific flux even when a highly turbid water with particles possessing the minimum back transport velocity was applied.
3. Fouling by both inorganic precipitates and particulate matter (a clay material) shows a negligible effect on reduction of flux in membranes. A substantial amount of solid deposits were observed by the SEM images from fouled membranes. However, the

results of the SEM combined with the flux experiments clearly showed that a layer with only inorganic particles created by either precipitates from softening or a clay material exhibited no resistance to the water flow through the membrane pores.

4. Organic matter, either natural organic matter or a simple organic component used as a NOM surrogate, is the most detrimental foulant on the membrane process. The flux decline was significant when raw waters (without softening) were applied regardless of the water source, i.e., natural or synthetic organic waters with dextran and alginic acid.
5. Softening pretreatment significantly reduces organic fouling in ultrafiltration. For each water, increasing the extent of softening continually reduced the fouling, but the extent of the improvement of water flux varied depending on the feed water.
6. The cause of flux improvement in the experiments with the polysaccharides (dextran and alginic acid) likely stemmed from the preferential removal of its high molecular weight fraction in softening. However, Lake Austin water consisted of relatively low molecular weight organic matter; therefore, softening induced small changes in its molecular weight size distribution. Rather, changes in hydrophobicity with softening were closely associated with the flux improvements of Lake Austin water. Therefore, softening has an ability to reduce foulants prior to ultrafiltration by preferential removal of both the high molecular weight fraction and hydrophobic portions of NOM.
7. The combined fouling with kaolin and dextran showed little effect of particles (i.e., kaolin) since no difference from the fouling with only organic matter was observed. The similarity might stem from the fact that polysaccharide has no ability to be adsorbed on the clay material, so that the polysaccharide behaves the same with respect to membrane with or without clay material being present. The effects of fouling by natural particles, which are presumably coated with NOM, were shown with experiments with Missouri River water. The much greater flux decline with Missouri River water than Lake Austin water indicates the importance of particle fouling since the big difference between Lake Austin water and Missouri River water is the turbidity; these waters have similar NOM concentration and characterization.
8. Softening also reduces fouling and improves the water flux in ultrafiltration membranes when a highly turbid water such as Missouri River water is treated. The improvement of flux was significant even with the standard and enhanced softening condition for Missouri River water.
9. Three cleaning methods applied at the end of each run effectively reveal different fouling mechanisms. Dextran apparently chemically attached to the membrane, since the surface wash showed little effect and the caustic wash showed a relatively significant effect on the recovery. Substantial amounts of alginic acid seemed to be deposited on the membrane surface without attachment since the surface wash recovered most of the lost flux. In addition, the relatively successful recovery by the acidic wash implies that alginic acid is bound with inorganic matter to some extent.
10. The SEM images of the surface of the membranes are quite valuable in understanding the nature of foulants. However, the SEM images sometimes fail to predict the reduction of flux; therefore, other measurements are necessary to investigate actual fouling phenomena in the membrane processes. The XPS was beneficial in determining the relative composition of each fouling material on the membrane surfaces.

CHAPTER 1
INTRODUCTION

Membrane processes, especially low-pressure membrane processes such as microfiltration (MF) and ultrafiltration (UF), are now frequently considered for application in drinking water treatment. In 1994, only one facility (San Jose Water Company, Saratoga, CA) in the United States used MF to treat river water. Now at least 60 facilities are in various stages of completion and have a combined capacity in excess of 300 million gallons per day (Freeman, Horsley, and Hess 2000). The significant increase in the use of membrane technology results from the recent changes in regulations that guide the water industry (*i.e.*, the Surface Water Treatment Rule (SWTR), the Interim Enhanced Surface Water Treatment Rule (IESWTR), and the Disinfection/Disinfectant By-Product Rule (D/DBP rule).

Due to these regulations, most drinking water systems have been forced to consider substantial changes in treatment processes, especially for microorganisms and natural organic matter (NOM). Although NOM itself is a naturally occurring material from decomposition of plants, it can react with chlorine or other disinfectants and create disinfectant by-products (DBPs). The DBPs are potentially hazardous to human health; for instance, trihalomethanes, one group of DBPs, include some cancer-causing materials. However, adequate disinfection of microorganisms is also necessary to prevent water-borne diseases. Therefore, a major challenge in drinking water systems is to balance the risks from microbial pathogens and disinfection by-products.

Low-pressure membrane processes have shown excellent microbial removal, in addition to turbidity and particle removal, regardless of pretreatment conditions. Furthermore, ultrafiltration (at least with the tightest membranes available) can accomplish some removal of NOM to meet those requirements prior to disinfection and, because of the NOM removal, reduce the subsequent DBP formation to acceptable levels.

1.1 PROBLEM STATEMENT

The biggest impediment for applying membrane processes is fouling that comes from mass flux (such as particle and organic matter) to the membrane surface and its pores due to convection flow through the membrane. The flow resistance increases as a membrane becomes fouled, so the membrane has to be cleaned periodically. Therefore, central issues for the further development and use of membrane processes are to understand the fouling mechanisms and to minimize fouling.

NOM has been recognized as a main foulant in membrane processes; however, Champlin (2000) indicated that the degree of fouling by either NOM or particles depends on how fouling occurs in specific conditions. In addition, Chellam *et al.* (1997) observed that the particles determined the properties of fouling during nanofiltration with conventional pretreatment. Therefore, both NOM and particle fouling need to be examined to better understand the fouling phenomena in ultrafiltration.

Pretreatment before the membrane process has become an important aspect of membrane operations, because pretreatment can reduce the concentration or change the nature of foulants in the feed flow before they deteriorate the membrane performance. Although considerable research has been devoted to understanding fouling in membranes, less attention has been paid to the effect of pretreatment on fouling. Limited studies on pretreatment schemes for ultrafiltration have included coagulation followed by sedimentation or dissolved air flotation, carbon adsorption, and membrane processes with larger pore sizes.

Coagulation with sedimentation has been used as a primary process to remove particulate matter, including microorganisms, in drinking water systems for many years. In the last 20 years, studies have shown that coagulation processes substantially remove NOM and DBP precursors. Therefore, coagulation has been investigated as a pretreatment for membrane processes because of its capability to remove the main foulants, *i.e.*, NOM and particles, as well as the ability to use the existing infrastructure in many water treatment plants.

Softening is traditionally designed to remove hardness ions in hard waters but it also has capability to remove particles and organic matter. However, virtually no effort has been directed toward using softening as a pretreatment before membrane processes. This research was designed to fill that gap.

1.2 OBJECTIVES

Many utilities throughout the central U.S. and Florida use precipitative softening, and most of those plants require modifications to meet the recent changes in the regulations. Membrane processes are being considered by many of these softening utilities as a method to satisfy those regulations. Therefore, the objectives of the proposed research are as follows:

1. To understand the nature of the fouling mechanisms for ultrafiltration membranes when used for waters that either require softening or have been softened, and
2. To use that understanding to determine promising options for the use of precipitative softening as a pretreatment before ultrafiltration.

1.3 APPROACH

To achieve the objectives of the research, bench-scale softening jar tests and membrane filtration were involved. The source waters included two natural waters and synthetic waters that were designed to allow testing of the effects of specific constituents on both softening and membrane performance. The synthetic waters included one with inorganic constituents only (simulated to match Lake Austin water), waters with simple organic components, and a water with kaolin. The synthetic inorganic water and the water with kaolin were used to investigate inorganic and particle fouling without NOM effects on fouling. The synthetic organic waters were produced with simple organic components, *i.e.*, polysaccharides, to obtain an understanding of NOM fouling. The natural waters included those used by the city of Austin, TX (Lake Austin) and St. Louis County, MO (Missouri River).

Several fouling mechanisms could be hypothesized in softening: inorganic fouling by continuous precipitative softening, organic fouling, particle fouling, and combined fouling by particle and organic matter. To examine the inorganic fouling, various scenarios for softening pretreatment were tested using the synthetic inorganic waters; the scenarios included lime softening alone, lime softening with subsequent pH adjustment, lime-soda ash softening, and lime-soda ash softening with subsequent pH adjustment. A specific softening process that achieved the least inorganic fouling on the membrane surface was then selected and used in subsequent experiments to further investigate the effects of NOM and particles on membrane performance.

Softening tests were performed with various lime doses using standard jar test conditions. Since the softening process is used as a pretreatment for ultrafiltration, the degree of softening was investigated thoroughly.

Ultrafiltration experiments were performed in cross-flow mode in a laboratory-scale flat sheet module. Fouling was monitored by measuring flux decline and transmembrane pressure (TMP). At the end of each experiment, some membranes were cleaned with three cleaning methods, and the clean water flux was measured after each cleaning to evaluate the extent of three fouling mechanisms: surface deposition, organic fouling, and inorganic fouling. Other fouled membranes were further investigated with scanning electron microscopy (SEM) or X-ray photoelectron spectroscopy (XPS) to understand fouling as it occurred adjacent to the membrane surface.

Chapter 2 is a literature review of softening chemistry, NOM removal by enhanced softening, several fouling mechanisms in membrane processes, and pretreatment options for ultrafiltration. All of the operational and analytical procedures are described in Chapter 3. The results of the raw water characterization and selection of the specific softening condition are presented in Chapter 4. The investigation of fouling mechanisms in the integrated system with softening and ultrafiltration are described in detail in Chapter 5. The final chapter, Chapter 6, summarizes the significance of the results and includes recommendations for future work in this area.

CHAPTER 2
LITERATURE REVIEW

2.1 SOFTENING

Softening by precipitation is a conventional water treatment process used to remove hardness ions, primarily calcium and magnesium. During softening, other undesirable substances such as suspended solids and natural organic matter (NOM) are also removed (Liao and Randtke 1985). Softening chemistry is discussed in this section, which also helps to understand NOM removal within softening, *i.e.*, enhanced softening.

Calcium is precipitated as calcium carbonate, $CaCO_3$, and magnesium is removed as magnesium hydroxide, $Mg(OH)_2$. The chemistry of Ca^{2+} and Mg^{2+} precipitation is expressed as shown below (Snoeyink and Jenkins, 1980).

$$CaCO_{3\,(s)} \leftrightarrow Ca^{2+} + CO_3^{2-} \qquad K_{S0,\,Ca} = [Ca^{2+}][CO_3^{2-}] = 10^{-8.48} \qquad (2.1)$$

$$Mg(OH)_{2\,(s)} \leftrightarrow Mg^{2+} + 2\,OH^- \qquad K_{S0,\,Mg} = [Mg^{2+}][OH^-]^2 = 10^{-11.16} \qquad (2.2)$$

As suggested by Equations 2.1 and 2.2, a higher pH than that found in natural water is required to start precipitation, so a base is added to induce the softening. Lime (CaO) is the most common base used to increase the pH in softening plants; when lime alone is added to the waters, the process is referred to as lime softening.

The chemistry of the associated anions, i.e., carbonate (CO_3^{2-}) and hydroxide (OH^-), is also included in the equilibrium description of softening as expressed in Equations 2.3, 2.4, and 2.5.

$$2\,H_2O = H_3O^+ + OH^- \qquad K_W = 10^{-14} \qquad (2.3)$$

$$H_2CO_3 + H_2O = HCO_3^- + H_3O^+ \qquad K_1 = 10^{-6.35} \qquad (2.4)$$

$$HCO_3^- + H_2O = CO_3^{2-} + H_3O^+ \qquad K_2 = 10^{-10.33} \qquad (2.5)$$

The total carbonate concentration (C_T), defined in Equation 2.6, and pH determine how much CO_3^{2-} and OH^- are available to react with the Ca^{2+} and Mg^{2+} in raw water, i.e.,

$$C_T = [H_2CO_3] + [HCO_3^-] + [CO_3^{2-}] \qquad (2.6)$$

For instance, if two raw waters have the same concentrations of Ca^{2+} and Mg^{2+} but different concentrations of total carbonate, the amount of $CaCO_3$ precipitate formed would be different.

Another important parameter, alkalinity, is introduced to account for many other species in raw water, which are not directly involved in the chemistry described in Equations 2.1- 2.5. Alkalinity reflects the capacity of water to neutralize added acids and can be expressed in different ways. Specifically, alkalinity is the amount of acid (in eq/L) that must be added to a water to bring that water to a proton reference state where the carbonate system is equivalent to a solution of carbonic acid. Several weak acid/base systems can contribute to the alkalinity. When the carbonate system (the dominant weak acid/base system in nature) is the only weak acid/base system of any significance, alkalinity can be simply expressed as shown in Equation 2.7.

$$\text{Alkalinity (eq/L)} = 2\,[CO_3^{2-}] + [HCO_3^-] + [OH^-] - [H_3O^+] \quad (2.7)$$

In addition, effects of other strong cations (K^+ and Na^+), strong anions (SO_4^{2-} and Cl^-), and other weak acid/base systems (PO_4^{3-} and SiO_3^{2-}) can be evaluated with alkalinity. Using the definition, it is possible to show that, when carbonate is the only weak acid/base system present, alkalinity can also be expressed as the difference between the strong base cations and the strong acid anions, with all constituents expressed in eq/L, *i.e.*,

$$\begin{aligned}\text{Alkalinity (eq/L)} = {} & 2\,[Ca^{2+}] + 2\,[Mg^{2+}] + [K^+] + [Na^+] + \{\text{other strong base cations}\} \\ & - 2\,[SO_4^{2-}] - [Cl^-] - \{\text{other strong acid anions}\} \quad (2.8)\end{aligned}$$

In precipitative softening, the calcium and magnesium concentrations are changed substantially but the other strong acids and bases remain unchanged (unless sodium carbonate is added), so one can lump the sodium and potassium terms into the "other strong base cations" and the sulfate and chloride into the "other strong acid anions" term, yielding the following:

$$\begin{aligned}\text{Alkalinity (eq/L)} = {} & 2\,[Ca^{2+}] + 2\,[Mg^{2+}] + [\{\text{other strong base cations}\} \\ & - \{\text{other strong acid anions}\}] \quad (2.9)\end{aligned}$$

Equation 2.9 makes it clear that the progression of softening (i.e., changes in the concentrations of the hardness ions, Ca^{2+} and Mg^{2+}), can be followed by monitoring the alkalinity. The addition of lime in softening increases the calcium concentration and pH, and it increases the carbonate ion and hydroxide ion concentrations. As implied by Equation 2.1, calcium carbonate starts to precipitate at the calcium concentration at which the product of calcium and carbonate ions exceeds the solubility product. Therefore, with further addition of lime, the calcium concentration in the solution decreases until the carbonate ion becomes limited. After that point, further addition of lime increases the calcium concentration. Therefore, a certain dose of lime yields the minimal calcium concentration, which represents the optimal dose for calcium removal.

For most operating softening plants, the softening achieved at this point is sufficient, and therefore this dose is referred to herein as the optimal dose for softening alone. Magnesium hydroxide is precipitated at higher pH values (approximately pH 11) than $CaCO_3$. As increasing lime doses are added, the Mg^{2+} concentration decreases very slightly up to the point where the solubility product of $Mg(OH)_2$ is exceeded, and then decreases rapidly to a concentration near zero. Although the starting points for the precipitation of $CaCO_3$ and $Mg(OH)_2$ vary with concentrations of individual compounds in raw water, $Mg(OH)_2$ precipitation begins, in most natural waters, around or after the minimal point of Ca^{2+} concentration. Therefore, alkalinity could be relatively constant or could increase with increasing lime doses after the optimal point for Ca^{2+} precipitation, depending on the point at which Mg^{2+} precipitation starts.

In cases where enhanced softening is considered, $Mg(OH)_2$ precipitation is sometimes induced to obtain the required NOM removal. As discussed subsequently, the $Mg(OH)_2$ solids are thought to be more effective to remove NOM due to their fluffy shapes (i.e., higher surface areas) and positive charges on the surface. However, in the past, the high lime dose required for Mg^{2+} removal was generally avoided in practice because precipitation of $Mg(OH)_2$ led to voluminous sludge and substantial operational problems in cases where the sedimentation tanks had been designed to capture the more dense $CaCO_3$ solids. For most Midwestern waters, which have a substantial amount of magnesium, the optimal lime dose for softening alone is around the point at which calcium concentration was the least, but prior to magnesium hydroxide precipitation. However, optimal doses of softening process could be determined differently depending on the treatment goal of the systems (Ralls 1999). Although $Mg(OH)_2$ precipitation is rare in real systems, it is essential in this research to include conditions of $Mg(OH)_2$ precipitation to understand its role in NOM removal and to evaluate the effects on subsequent processes, *i.e.*, ultrafiltration.

2.2 NATURAL ORGANIC MATTER

Natural organic matter (NOM) has many different properties and is composed of an extremely complex mixture of compounds, most of which are not yet identified. However, the nature of NOM is somewhat distinguished by its origin: continents (i.e., pedogenic NOM) or oceans (i.e., aquagenic NOM).

2.2.1 Composition of NOM

Most drinking water resources, *i.e.*, rivers and lakes, contain different portions of both pedogenic and aquagenic organic matter depending on adjacent ecosystems. Pedogenic organic matter is created due to rain-water leaching of soil organic matter (SOM) (Buffle 1990). It is largely composed of fulvic acid with lesser amounts of humic acid, pedogenic polysaccharide, lipids, protein, and amino acids. After drainage from soil, hydrophobic macromolecules like humic acid, pedogenic polysaccharide, and lipids are likely to be retained in soils. Protein and amino acids are quickly degraded. Therefore, in most cases, the pedogenic fulvic acid remains and thus prevails in aquatic NOM (Wilkinson et al., 1997, Buffle et al., 1998). Aquagenic organic matter is mainly a result of excretion and decomposition of plankton and aquatic bacteria (Buffle 1990). It includes proteins and cell wall derivatives such as polysaccharides and peptidoglycans, which are high molecular weight carbohydrates.

Therefore, major NOM groups in most lakes and rivers are (1) humic substances (the pedogenic fulvic acids and some humic acids leached from soils), (2) rigid biopolymers (the aquagenic carbohydrates), (3) aquagenic refractory organic matter, and (4) aquagenic proteins, which are relatively biodegradable (Buffle 1990; Owen et al., 1995; Krasner et al. 1996).

2.2.2 Properties of Humic Substances

In general, humic substances from soil organic matter, i.e., fulvic acid and humic acid, make up more than 70% of aquatic NOM (Wilkinson et al., 1997). Humic substances are derived from the separation of soil organic matter. After extraction of soil, the insoluble organic material is humin, and the soluble material is the humic substances, which are further divided into humic acid and fulvic acid by acidification. The fulvic acid is the soluble fraction, and the humic acid is the insoluble fraction at low pH. The humic acids are less oxidized and more polymerized molecules than fulvic acids. Therefore, humic acid has higher molecular weight (approximately 10-30 kDa) than the fulvic acid (approximately 3-10 kDa) in natural water (Jucker and Clark 1994). In addition to high molecular weights, the humic acid has relatively low aqueous solubility due to the lesser amount of carboxylic acid; thus most natural waters contain five to twenty-five times more fulvic acid than humic acid. Furthermore, the humic acid is more hydrophobic because of the longer chains of fatty acids (Vik and Eikebrokk 1989).

2.2.3 Fractionation of NOM

Various fractionation methods are used to classify NOM into homologous compounds due to difficulties in identification of each NOM compound (Owen et al., 1995, Krasner et al., 1996). Three properties are frequently used to separate NOM: size, hydrophobicity, and charge.

Separation by size is often performed by size exclusion chromatography or ultrafiltration. The separation by size corresponds fairly well to the major groups of NOM. For instance, about 70% of pedogenic refractory organic matter (i.e., mainly fulvic acid) has a size (measured as a molecular weight) between 300 Da and 10 kDa and polysaccharides are mostly larger than 160 kDa (Buffle 1990).

Second, the resin XAD-8 is used to separate NOM by differences in adsorption affinities, *i.e.*, hydrophobic or hydrophilic NOM (Collins, Amy, and King 1985; Nilson and DiGiano 1996; Carroll et al., 2000). The term, hydrophobic or hydrophilic, is used as a relative affinity. Therefore, fulvic acid is said to be more hydrophilic than humic acid, but more hydrophobic than proteins. In general, proteins, carbohydrates, free amino-acid, and phenols are said to be relatively hydrophilic (Buffle 1990). Furthermore, hydrophobicity has significant impacts on drinking water treatment processes; hydrophobic components are known to be more reactive and more easily removed.

Charge characteristics are also used to separate NOM using ion-exchange resins. Cation-exchange resins can purify a group of negatively charged homologous compounds from the inorganic cations retained on the resin. In addition, proteins can be separated from other organic compounds because they are cationic under acidic conditions while others are neutral or negatively charged (Buffle 1990).

In addition, analysis by gas chromatography (GC) – pyrolysis – mass spectrometry (MS) has been used to classify NOM into polysaccharide, polyhydroxyaromatics (PHA), protein, and amino-sugar fractions. The method has been used to identify the removal characteristics of each NOM component in coagulation and membrane process (Mallevialle et al. 1989; Krasner et al. 1996). For instance, the results showed that polysaccharide had a relatively low removal rate in coagulation, but might have the highest fouling ability in ultrafiltration (Lahoussine-Turcaud et al. 1990, Mackey and Wiesner 1999).

Careful consideration of the above separation methods reveals that no individual fractionation method has the capability to separate all the major groups of NOM from each other. Rather, detailed separation can only be achieved by a combination of two or more methods (Buffle 1990).

2.3 ENHANCED SOFTENING

Softening is traditionally designed to remove hardness ions in hard waters but it has also shown capabilities of substantial removal of organic matter. NOM removal in softening, *i.e.*, enhanced softening, might have significant effects on fouling in membrane processes. Therefore, removal mechanisms of NOM and several factors affecting NOM removal in softening are discussed in detail in this section.

2.3.1 Removal Mechanisms of NOM in Softening

Several mechanisms might play a role in the removal of organic contaminants by softening: (1) surface adsorption, (2) coprecipitation (occlusion and adsorption), (3) precipitation of insoluble organic compounds, (4) coagulation-flocculation, and (5) chemical reaction (Randtke et al. 1982). Among these mechanisms, coprecipitation has appeared to be the main mechanism.

Coprecipitation is defined as the contamination of a precipitate by an impurity that is otherwise soluble under the conditions of precipitation (Randtke 1988). Coprecipitation can occur in four different ways: occlusion, surface adsorption, isomorphic inclusion, and nonisomorphic inclusion. Occlusion and surface adsorption, which occur at lattice sites as the crystals are growing, play the major roles in coprecipitation. The last two inclusion mechanisms are not considered relevant here because NOM has substantially different size and chemical characteristics from the precipitating ions, so it is impossible to be included inside the precipitate lattice.

The coprecipitation by occlusion is especially effective in the early stage of rapid mixing, when calcium carbonate precipitates are amorphous. Images of the morphology of $CaCO_3$ taken by scanning electron microscopy (SEM) showed that the calcium carbonate precipitate was amorphous during the two minutes of rapid mixing, and then the $CaCO_3$ was rapidly transformed into the well-defined rhombohedral calcite crystals (Liao and Randtke 1985). Along with the differences in morphology, the specific surface areas of $CaCO_3$ in the two stages also changed from 15.00 m^2/g to 1.36 m^2/g. Thus the amorphous form is more reactive due to the high surface area.

Surface adsorption of NOM is the accumulation of NOM at an interface of a precipitate such as $CaCO_3$ and $Mg(OH)_2$ solids. However, only a minor amount of organic matter (less than 8.6% of peat fulvic acid) was removed through surface adsorption on the $CaCO_3$ solid (Randtke et al. 1982); therefore, NOM removal by surface adsorption onto $CaCO_3$ precipitate alone is unlikely to be a main removal mechanism in practice. Magnesium hydroxide is thought to be a better adsorbent of NOM due to a higher surface area than $CaCO_3$. However, most softening plants are operated without $Mg(OH)_2$ precipitation due to its voluminous sludge production.

Liao and Randtke (1985) also examined the removal of fulvic acid in lime softening. They found that the adsorption of ground water fulvic acid (GWFA) did not occur on $CaCO_3$ solids without calcium ions being in excess. Therefore, they postulated that the main mechanism to remove GWFA was adsorption of fulvic acid (or fulvate) onto $CaCO_3$ solids. Although both NOM and calcium carbonate precipitate were negatively charged in the softening condition, accumulated Ca^{2+} ions near the precipitate facilitated the adsorption by linking NOM to $CaCO_3$.

Other mechanisms for NOM removal such as precipitation of insoluble organic compounds, coagulation, and chemical reaction might have occurred under some very specific conditions but could not be the main mechanisms. For instance, precipitation can occur when the aqueous solubility of a compound is exceeded. Randkte *et al.* (1982) revealed that calcium fulvate was precipitated under the condition of no carbonate concentration in solution. However, it was doubtful that it could occur with the extensive presence of carbonate ions in practice.

Although several previous studies have examined the NOM removal mechanisms, little attention has been paid to investigate the effects of each softening precipitate on NOM removal. Since each precipitate has a different capability due to its properties, it is useful to scrutinize the crystallization of each precipitate in softening.

2.3.2 Crystallization in Softening

A variety of precipitates including calcium carbonate, magnesium hydroxide, and calcium phosphate can be formed in softening, depending on compounds in a source water. The characteristics of precipitates such as size, structure, and surface charge are very important to both softening and ultrafiltration. These properties substantially influence the removal of hardness ions and NOM because they affect settling velocity and the affinity for individual compounds. For instance, settling velocities of precipitates determine softening performance and might control the loading rate of particles in ultrafiltration if settling is applied.

Forms and Shapes of CaCO3

Calcium carbonate, a major precipitate in softening, has several different forms: calcite, aragonite, vaterite, $CaCO_3$ with attached waters of hydration ($CaCO_3{\cdot}H_2O$, $CaCO_3{\cdot}3H_2O$, and $CaCO_3{\cdot}6H_2O$), and amorphous $CaCO_3$. Calcite is usually the most prevalent form in softening, although the dominant form of $CaCO_3$ depends on solution conditions including organic matter concentrations (Randtke et al. 1982; Liao and Randtke 1985; Peters, Baumann, and Larson 1989).

Forms or shapes of $CaCO_3$ are changed with reaction time and determined by solution chemistry. Liao and Randtke (1985) found that $CaCO_3$ was amorphous at the initial stage of precipitation, and then transformed to a well-defined crystal. Their batch test without any other compounds showed that the transformation to the calcite crystal occurred within two minutes and the softening was almost finished within five minutes. However, the study by Alexander and McClanahan (1975) revealed that the apparent equilibrium level of the calcium ion concentration was reached after 25 minutes at either pH 10 or pH 12 and the rate of the calcium ion removal was greatest at the beginning and then slowly decreased with time. Therefore, they concluded that kinetics of the precipitation of $CaCO_3$ was first order with respect to the calcium ion concentration. Nancollas and Reddy (1974) also indicated a slow reaction under the super-saturation of calcium ion, in which the precipitation was completed within approximately 20 minutes. However, they suggested that the precipitation was a second order reaction with respect to the calcium ion concentration. The faster kinetics by Liao and Randtke might be due in part to the higher initial concentration (6 mM of Ca^{2+}) and the relatively high initial pH (pH 11) at the beginning in comparison to the others (i.e., 2.5~ 5 mM Ca^{2+} and pH 9~12 or 6-7 mM Ca^{2+} and pH 9). The study by Alexander and McClanahan (1975) proved that increasing pH increased the kinetics of $CaCO_3$ precipitation.

In addition to kinetics, Nancollas and Reddy (1974) showed significant effects of different seed materials on crystal shapes. The images from SEM of calcite revealed that the specific surface areas varied from 0.35 to 6.2 m^2/g depending on the different seed materials in the solutions. The result implies that the shapes of the precipitates might be different from source to source due to different water characteristics, especially colloidal particles that can be used as seed materials.

Effects of Nucleation and Crystal Growth

Physical properties of precipitates such as size and shape are also dictated by nucleation and crystal growth pattern. Randtke et al., (1982) illustrated that crystallization reactions were different depending on the base used to increase pH in softening. Three bases are commonly considered in practice: lime (CaO or $Ca(OH)_2$), soda-ash (Na_2CO_3), and caustic soda ($NaOH$). Although lime is the most popular base in softening, soda-ash is used in addition to lime for waters in which the carbonate concentration is too low to accomplish a proper softening. However, caustic soda is seldom used for softening due to the poor softening performance. Lime softening produced a few nuclei which grew slowly, thus leading to well-structured and dense calcite crystals. The crystals induced by lime were small and settled slowly. This led to the high removal of calcium ions in the solution. In contrast, softening initiated by NaOH addition produced many nuclei which were precipitated very fast and yielded a relatively high residual calcium concentration. The soda-induced crystals were metastable forms of $CaCO_3$, mostly vaterite and a small amount of calcite.

In addition to the low cost, the greater removal of calcium by lime is the reason for its frequent use in softening. It should be noted also that few, if any, plants would ever rely solely on caustic soda to induce precipitation because of the poor softening and the high cost.

Effects of Impurities

The structure of $CaCO_3$ is also altered by coprecipitated compounds, i.e., impurities in the crystal lattice such as NOM and magnesium. In general, NOM is adsorbed on the surface of inorganic particles and thus stabilizes the inorganic particles in aquatic systems (Buffle 1990). The presence of fulvic compounds stabilized ferric hydroxide for more than 5 months in a fresh water by hindering the crystal growth. In addition, Randtke et al., (1982) showed that the residual calcium concentration in the final solutions increased as the initial TOC concentration was increased, which implied the hindrance of $CaCO_3$ formation by peat fulvic acid. Nancollas and Reddy (1974) also illustrated that a small amount of phosphorous-containing anionic additives or phosphonate salt, which has been used for scale control in drinking water systems, markedly inhibited the rate of crystallization. Finally, magnesium ion is known to inhibit the $CaCO_3$ precipitates (Reddy and Wang 1980).

It is interesting to note that the $CaCO_3$ crystal inhibition by a compound relates to the removal of NOM by $CaCO_3$ precipitation. For instance, the presence of magnesium or phosphate ion is known to inhibit $CaCO_3$ growth as well as to enhance NOM removal.

Importance of Mg(OH)2 Precipitates

Magnesium hydroxide as a precipitate in softening becomes more interesting since softening has been considered to remove NOM prior to chlorine disinfection for controlling DBP precursors. Precipitates of $Mg(OH)_2$ have been examined in many applications, including industries such as textile and Kraft Mill industries (Judkins and Hornsby 1978, Oldham and Rush 1978) and water and wastewater systems such as dewatering and thickening processes (Peters, Baumann, and Larson 1989). $Mg(OH)_2$ precipitates have been also used for industrial wastewater containing high color contents and chemical oxygen demand (COD) because magnesium is able to be removed from precipitated sludge and recycled through the processes. $Mg(OH)_2$ precipitates have been shown to cause higher NOM removal than $CaCO_3$. For instance, the COD removals of total Kraft Mill effluent using three different regimes (i.e., lime alone, lime with 30 mg/L Mg^{2+}, and lime with 60 mg/L Mg^{2+} addition) were respectively 70%, 85%, and 93% (Oldham and Rush 1978).

The higher removal of NOM by $Mg(OH)_2$ has been explained by its petal-like and gelatinous, or amorphous, shape, thus yielding a high surface area. In addition, a positive surface charge of $Mg(OH)_2$ under softening conditions has rendered the possibility of high NOM removal (Randtke et al. 1982; Peters, Baumann, and Larson 1989; Fradin and Field 1999).

However, as mentioned earlier, $Mg(OH)_2$ solids are usually avoided in softening because of the voluminous sludge production, which can cause severe problems in dewatering and thickening (Peters, Baumann, and Larson 1989). In addition to $Mg(OH)_2$ precipitation, Mg^{2+} ion coprecipitation into $CaCO_3$ causes sludge with the molar ratio of calcium to magnesium less than two to be difficult to dewater.

Summary

In this section, the properties of two major precipitates, $CaCO_3$ and $Mg(OH)_2$, in softening have been analyzed. They have differences in the required pH to initiate precipitation, the shapes and surface charges in precipitates, and the extent of NOM removal. Nevertheless, they also have similar properties such as the ability to adsorb NOM and the crystallization inhibition by NOM. However, little research has been performed to characterize the crystallization of $Mg(OH)_2$ and other precipitates besides $CaCO_3$ in softening, although the importance of $Mg(OH)_2$ precipitate has been recognized. Effects of different precipitates need to be thoroughly investigated under various conditions in softening and subsequent processes.

2.3.3 NOM Removal in Softening

Softening can cause substantial removal of natural organic matter including humic acid, fulvic acid, specific organic contaminants, and DBP precursors. Several factors affect NOM removal in softening: pH, initial organic concentration, hydrophobicity, and molecular weight distribution. These factors are explained below.

Effects of pH

In general, DOC removal increases as pH increases. Liao and Randtke (1985) found that removal of ground water fulvic acid increased from 28% to 44% as pH was increased from pH 9 to pH 12 with carbonate ion in excess. Since they added no magnesium ions into solution, the increased removal related to the increased precipitation of calcium carbonate and perhaps a change in the NOM characteristics and the surface charge of $CaCO_3$ at the high pH. In addition, Thompson et al., (1997) explained the increased DOC removal with pH and Mg precipitates. They observed that the amount of DOC removal per unit change in pH increased significantly after the pH at which Mg precipitation occurred extensively. In this case, increasing pH accompanied with $Mg(OH)_2$ precipitation dramatically increased DOC removal.

The research by Ralls (1999) and Smith (2001) with natural water sources also showed that DOC removal increased with increasing lime doses (i.e., increasing pH). However, the results from five water sources, representing a broad spectrum of hardness and NOM conditions in hard waters throughout the country, indicated that substantial NOM removals could be achieved in some waters with relatively high DOC and SUVA concentration even before substantial $Mg(OH)_2$ precipitation occurred (Smith et al., 2002).

Effects of Initial Organic Concentrations

The initial DOC concentration in raw water also affects overall removal of DOC in softening. Randtke et al., (1982) used peat fulvic acid to investigate DOC removal by softening as a function of initial TOC concentration. As the TOC concentration of the peat fulvic acid was decreased from 10 to 0.5 mg/L, the TOC removal of the peat fulvic acid with a constant dose of lime increased from 43% to 95%, but the final calcium concentration remaining in the softened water decreased from 210 to almost 0 mg/L as $CaCO_3$. The results showed that as the initial TOC concentration decreased, the fractional removal of TOC increased, but the absolute amount of TOC removed decreased. Also, the results indicated that the high TOC concentration inhibited the removal of calcium ions by interfering with $CaCO_3$ precipitation.

Thompson et al., (1997) investigated the DOC removal in softening with nine natural water sources with different DOC concentration. In this study, both $CaCO_3$ and $Mg(OH)_2$ were simultaneously precipitated because real waters containing Ca and Mg were used. The DOC removal for these nine waters showed that the softening process could substantially remove NOM with small amounts of Mg precipitate, although it was least effective for the source water with the lowest TOC concentration, i.e., 2.5 mg/L of TOC in Austin water. Note that this result is consistent with the results found by Randtke et al., (1982) based on the absolute amount of TOC removed.

In addition, the research by Smith et al. (2002) with five natural water sources indicated that DOC removal was substantial in the waters with high initial DOC concentration. The least DOC removal occurred in a water with the lowest DOC concentration and the low Ca^{2+} raw concentration, leading to the low solids production and limited adsorption sites.

Therefore, softening either with $CaCO_3$ precipitation alone or with some $Mg(OH)_2$ precipitation would be expected to remove a greater absolute amount of TOC for a source water with a high initial TOC concentration than a source water with a lower initial TOC concentration.

Effects of Hydrophobicity

The study by Liao and Randtke (1986) indicated that molecular characteristics, including molecular charge, functionality, and affinity for water influenced NOM removal. The results from the speciation of groundwater humic substances from hydrophobic to hydrophilic, i.e., humic acid, fulvic/humic acid, and fulvic acid showed that the removal of TOC was, respectively, 46%, 40%, and 34%, which implied that hydrophobic compounds were more easily removed in the softening process. In their other study (Liao and Randtke 1985), the authors indicated that the compounds least soluble with calcium ions would be most strongly adsorbed and coprecipitated. Therefore, as hydrophobicity is increased, the removal of the compounds is increased because there is a means of adsorption onto the $CaCO_3$ surface.

Using nine natural waters with different hydrophobic DOC concentrations, Thompson et al., (1997) investigated effects of Mg precipitation on the DOC removal. The Mg precipitates were the most effective in removing NOM from waters with the greatest portion of hydrophobic DOC (53%) and the least effective to remove NOM from waters with the lowest portion of hydrophobic DOC (34%). A significant correlation between hydrophobic DOC and Mg removals implied that the amount of DOC removal by $Mg(OH)_2$ would increase as hydrophobic DOC in the source water increased.

The results by Smith et al., (2002) also indicated that softening was most effective in removing the reactive and hydrophobic portion of NOM, which was shown by a decrease in the SUVA with increasing lime doses.

Effects of Molecular Weight Distributions

Organic matter with high molecular weights has been preferentially removed in water treatment processes such as softening and coagulation (Liao and Randtke 1986, Semmens and Staples 1986, Tambo and Kamei 1989). Liao and Rantdke (1986) indicated that organic matter containing numerous carboxyl and phenolic functional groups, such as humic substances and proteins, was better removed because of substantial dissociation of the functional groups under high pH conditions, i.e., softening. The dissociation rendered greater binding sites for Ca^{2+}, so that the organic matter easily adsorbed or precipitated. Also, polymers with high molecular weights were more likely to be removed through complexation and adsorption than their monomeric analogs. Even a polymer with a low bonding energy per segment could be strongly adsorbed on $CaCO_3$ due to many contacts of the polymer segments.

Effects of Chemical Addition

Effects of several common chemical additions and NOM removal in softening using six natural waters (two from one source) were investigated by Ralls (1999) and Smith (2001). The results showed that enhanced softening improved DOC removal, decreased the specific UV absorbance at 254 nm (SUVA), and decreased dissolved organic halogen formation. The results from chemical addition also found that softening performance may be strongly influenced by the raw water bromide concentrations to control brominated disinfection by-products (DBPs). In addition, softening plants treating waters with high silicate concentration might have difficulty achieving the desired TOC reduction. Apparently, Mg-Si precipitates are formed that are less effective than $Mg(OH)_2$ in removing NOM.

2.4 ULTRAFILTRATION

2.4.1 Classification of Membrane Processes

Membrane processes have been widely used in a variety of industries. Reverse osmosis (RO) has been used for desalination of seawater for years. In the food industry, especially in the beverage industry, ultrafiltration or nanofiltration is a very popular process to concentrate fruit juice. Recently, nanofiltration has been used as an alternative method of softening and has been extensively applied in Florida (Bergman 1996). In addition, nanofiltration has been shown to remove DBP precursors in drinking water systems (Wilkes et al., 1996, Chellam et al., 1997).

Low-pressure membrane processes such as microfiltration (MF) and ultrafiltration (UF) are increasingly being applied to drinking water systems. MF and UF have been recognized as alternative filtration processes for removing microorganisms from drinking water to meet regulations (Jacangelo et al., 1991; Karimi, Vicker, and Harasick 1999, California DHS, 2001). To help determine compliance with the SWTR easily, the State of California Department of Health Services initiated a project to certify the log-removal credit for Cryptosporidium, Giardia, and viruses for various membrane systems. A draft report (California DHS, 2001) indicated that all four ultrafiltration membranes tested were granted a 4-log removal credit for Giardia and Cryptosporidium, and two of them were also given 4-log credit for viruses. The other two systems were credited with 2 and 3.5 log removal credit for viruses.

The membrane classification as RO, NF, UF and MF depends on their pore size. Ranges of typical characteristics of each membrane process are presented in Table 2.1.

2.4.2 Application of Low-pressure Membrane Processes

Low-pressure membrane processes in drinking water systems offer many benefits. The benefits include reliable finished water quality regardless of the quality of raw waters (without chemical additions), an automated control system (which reduces operator costs), a small land acquisition, and long-term compliance with regulations (Cleveland 1999; Freeman, Horsley, and Hess 2000). For instance, when the city of Marquette, Michigan, needed to filter the Lake Superior water supply, the city selected MF rather than a direct deep bed filtration system. The choice stemmed from the fact that it met the disinfection criteria (CT rule) in the SWTR and required only half of the available area (Kelley and Olson 1999).

Ultrafiltration appears to merit particular interest in the drinking water industry. With recent advances in the membrane industry, ultrafiltration can accomplish excellent disinfection by removal of microorganisms and some removal of NOM to meet those requirements prior to disinfection. Because of the NOM removal, ultrafiltration can sometimes reduce the subsequent DBP formation to acceptable levels (Jacangelo et al., 1991).

Table 2.1
Classification and characteristics of membranes

Process	MWCO (Daltons)	Pore size (μm)	TMP (kPa)	Rejected Contaminant
Microfiltration	>500k	0.05-10	30-300	Large particles, bacteria
Ultrafiltration	1k-1000k	0.001-0.1	50-700	Colloidal particles, viruses, enzymes
Nanofiltration	50-200	NA	350-1000	NOM, protein, hardness
Reverse osmosis	1-100	NA	800-8000	Salts, metal ions, small organic molecules

Adapted from Chang (1996) and Wiesner and Aptel (1996)

Further, ultrafiltration can accomplish these removals at reasonable pressures and recoveries. For small plants, ultrafiltration can be used as the sole treatment prior to adding a disinfectant for the distribution system; a particular value for small plants is that the operation can be automated and monitored remotely. The costs are also competitive with conventional treatment for small systems, because the number of unit processes and need for chemical feeders are much greater in conventional systems (Wiesner et al., 1994). For larger systems, however, where the economies of scale for several processes are greater than they are for membrane processes, and where the availability of trained personnel is better, membrane processes make more sense as part of a treatment train using conventional processes as pretreatment to reduce the load on the membrane system.

2.4.3 Membrane Materials

Several polymers and other materials are used for manufacturing membranes. Popular materials in ultrafiltration include cellulose acetate and its derivatives and polysulfone (Freeman, Horsley, and Hess 2000). Structures of these membranes are presented in Figure 2.1. A typical structure of a completely substituted molecule of cellulose, i.e., cellulose triacetate, is shown in Figure 2.1(a). Traditionally, commercial products in practice are less than completely substituted; thus, the degree of substitution is approximately 2.4-2.5 instead of being 3.0 in cellulose triacetate. Cellulose and its derivatives are generally linear, rod-like, and rather inflexible molecules (Cheryan 1986). Some advantages of these materials are relatively high flux, high salt rejection properties, and less fouling tendency because of its hydrophilic property. However, there are also disadvantages such as a narrow range of applicable temperature and pH and its poor resistance to chlorine.

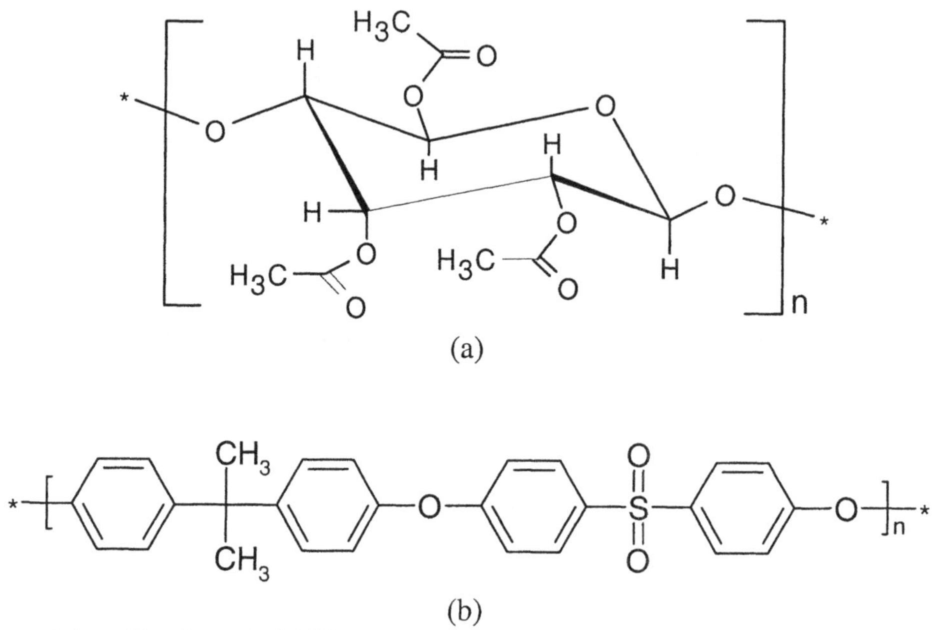

(*Source:* Adapted from Pontie et al. 1997)

Figure 2.1 Typical structures of (a) cellulose triacetate and (b) polysulfone

Table 2.2
Relative hydrophobicity of membrane materials

Reference	Ordered from Hydrophilic to Hydrophobic							
Laine et al., 1989	CA				PS			
Jucker and Clark 1994	CA	PA			PS			
Membrex (Ind.)				PS	PAN	PVDF	PP	PTFE
Lloyd 1999	CA	PA	PAN	PES	PS	PE/PP	PVDF	PTFE

Abbreviations: CA (cellulose acetate), PA (polyamide), PAN (polyacrylonitrile), PES (polyethersulfone), PS (polysulfone), PVDF (polyvinylidene flouride), PE (polyethylene), PP (polypropylene), PTFE (polytetrafluoroethylene)

Table 2.3
Contact angle of clean ultrafiltration membranes

Membrane	Contact angle	Material	MWCO	Reference
Amicon PM30	42.5	Polysulfone	30kD	Jucker and Clark (1994)
Amicon XM50	40	Poly(acrylonitrile-co-vinyl chloride)	50kD	
Amicon YM10	6.5	Cellulose acetate	10kD	
Amicon YM30	7.0		30kD	
Amicon YM1	96		1kD	
Amicon YM100	31		100kD	
Amicon PM10	38.4	Polysulfone	10kD	Oldani and Schock (1989)
Amicon PM30	43		30kD	
DDS GR61/PP	44.3		20kD	
Desal CA-UF	45.9	Cellulose acetate	5.5kD	
Desal E100	56.5	Polyethersulfone	25kD	
Kalle UF PS 15/PP60	39.5	Polysulfone	15kD	

Polysulfone has a strong and rigid structure characterized by repeating units of diphenylene sulfone as shown in Figure 2.1(b). The structure allows a wide range limit of temperature and pH and has fairly good tolerance to chlorine. However, polysulfone shows a relatively great tendency to foul because of its high hydrophobic property.

A hydrophobic membrane has a relatively greater affinity for NOM than feed water. Table 2.2 shows the relative hydrophobicity of various membrane materials. In general, cellulose acetate (CA) is the most hydrophilic membrane available. Laine et al., (1989) illustrated that greater fouling was found using a hydrophobic membrane than a hydrophilic membrane. This result yielded the development of more hydrophilic materials by membrane manufacturers.

The hydrophobicity of membrane surface is usually measured by the contact angle, which technically analyzes the interaction of three phases, i.e., membrane, water, and air (Jucker and Clark 1994; Bouchard, Jolicoeur, and Kouasio 1997). The higher the contact angle between water and the membrane surface, the greater the hydrophobicity of the membrane. Table 2.3 summarizes contact angle measurements of ultrafilters with different membrane materials. In general, cellulose acetate has a small contact angle, which is consistent with its hydrophilic properties. The contact angle of YM1 showed an exceptionally high value of the contact angle; the author implied that its rough surface caused the high value of contact angle. In addition, the relatively great contact angle of another cellulose acetate membrane, Desal CA-UF, could be influenced by the smaller pore size (*i.e.*, 5.5kD of MWCO). Polysulfone and polyethersulfone have higher values of contact angle, which is expected from their hydrophobic properties. Contact angle analysis, furthermore, can provide insights about adsorption tendency for humic substances on a membrane surface. Jucker and Clark (1994) showed that the contact angle decreased as humic substances were increasingly adsorbed at pH 6. Because humic substances were hydrophilic at pH 6, the membrane surface became more hydrophilic by adsorption of humic substances (*i.e.*, coating the membrane with these hydrophilic compounds).

2.4.4 Membrane Performance

In general, membrane performance is evaluated with two parameters: filtrate flux and rejection.

Filtrate Flux

Filtrate flux is defined as a water production rate per unit area of membrane. Two approaches are often used to predict the filtrate flux: the standard film theory and the thin film theory.

The standard film theory is derived from Darcy's law, which predicts filtrate flux based on hydraulic resistances from the membrane itself and from fouling layers (Chang 1996, Wiesner and Aptel 1996). The flux of clean water through a membrane without any resistance from deposited fouling layers is often described as follows:

$$J = \frac{\Delta p - (\sigma_k \cdot \Delta \Pi)}{\mu \cdot R_m} \qquad (2.10)$$

where Δp is the transmembrane pressure drop (TMP, kPa), σ_k is treated as an empirical constant, $\Delta\Pi$ is the change in osmotic pressure across the membrane (kPa), μ is the absolute viscosity of the water (1 centipoise = 0.01 g/cm-s at 20°C), and R_m is the hydraulic resistance of the clean membrane (1/m). In ultrafiltration, the differential osmotic pressure is usually negligible since ultrafiltration rejects macromolecular and colloidal species, which exhibit usually quite small values of osmotic pressure. Sometimes, Δp_{net} is used to account for the net transmembrane pressure drop including effects of osmotic pressure (Wiesner and Aptel 1996).

As fouling materials accumulate in the membrane through deposition on the membrane surface and adsorption to pore walls, resistances of fouling layers increase over time. Therefore, the reduced filtrate flux caused by fouling materials is expressed with a series of resistances as follow:

$$J = \frac{\Delta p_{net}}{\mu(R_m + R_c + R_{cp} + R_a)} \tag{2.11}$$

where R_c, R_{cp}, and R_a are hydraulic resistances due to a cake layer, a concentration polarization layer, and solute adsorption in the membrane pore, respectively.

The standard film theory describes the relationship between transmembrane pressure and filtrate flux. In addition, the theory could provide a method for estimating the cake and membrane resistances as a function of water production and experimental conditions.

The thin film theory is used to predict filtrate flux when the flux is limited by mass transfer of the solute and is independent of pressure. The theory also assumes that advection and diffusion are the only mechanisms for solute transport and the filtrate flux is at the steady state condition (Chang 1996).

As fouling materials accumulate near a membrane over time, concentrations of solutes are increased near the membrane surface, a phenomenon called concentration polarization. The thickness of the concentration polarization layer is a function of the relative magnitude of advective flow by filtration and diffusive transport of the solutes, as illustrated in Figure 2.2 (Wiesner and Aptel 1996). The mass balance in the boundary layer can be expressed as follows:

$$J \cdot C = J \cdot C_p - D\frac{dC}{dx} \tag{2.12}$$

where J is the filtrate flux, C is the concentration of solute, C_p is the concentration in the filtrate, D is the diffusion coefficient of the solute, and dC/dx is the concentration gradient between the bulk and the surface. The boundary of the concentration polarization layer is from the edge of the bulk flow to the membrane surface. Integration of the above equation yields the expression:

$$J = \frac{D}{\delta} \cdot \ln\left(\frac{C_w - C_p}{C_b - C_p}\right) \tag{2.13}$$

where δ is the thickness of the concentration polarization layer (and thus D/δ is a mass transfer coefficient, k), and C_b, C_w, and C_p are the concentrations in the bulk, at the wall, and in the filtrate, respectively.

When the concentration of solute in the filtrate is negligible, the equation expresses the limiting filtrate flux as a function of the bulk concentration, the limiting wall concentration, and the mass transfer coefficient. This so-called film layer model describes filtrate flux under mass-transfer limited conditions (Wiesner and Aptel 1996).

$$J = k \cdot \ln\left(\frac{C_{wall}}{C_{bulk}}\right) \tag{2.14}$$

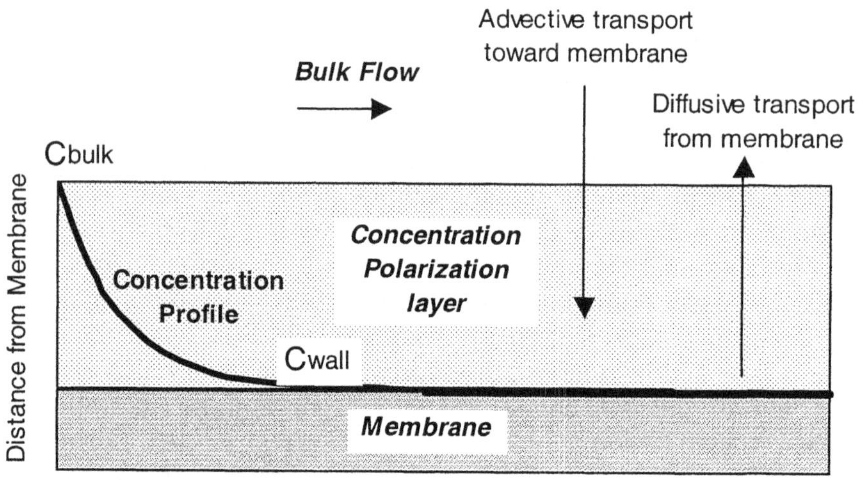

(*Source:* Adapted from Wiesner and Aptel 1996)

Figure 2.2 Concentration profile of concentration polarization layer

2.5 FOULING PHENOMENA

Membrane processes are mainly a physical treatment, in which particles, colloidal matter, or molecules larger than the membrane pores are removed by physical sieve mechanisms of the membrane. Therefore, materials removed from the water flow are deposited and accumulated on the surface of membranes. The deposition initiates membrane fouling.

Fouling is defined as reduction of water permeability due to the increasing resistances by deposition or adsorption of particles and organic matter (Wiesner and Aptel 1996). Several factors affect fouling phenomena: feed water characteristics, membrane characteristics, and operational conditions. Operational conditions such as transmembrane pressure and flow mode in the membrane module (i.e., dead end or crossflow mode) have effects on both fouling and filtrate rate (Crozes et al., 1997). The crossflow mode is designed to reduce accumulation on membrane surface by fluid dynamics. In the dead-end mode, feed water flows perpendicular to membrane surface. In comparison, the crossflow mode creates a tangential flow across membrane surface, which induces a shear force inside of the filtration system, thus reducing accumulation. The important feed water characteristics that influence fouling are pH, ionic strength, and the concentrations and nature of colloidal particles and NOM. Also, the membrane characteristics such as pore size, pore density, and hydrophobicity of membrane materials impact the extent of fouling.

Many models exist to explain fouling phenomena in membrane processes. Consideration of mass transport of colloidal suspensions is useful to estimate fouling by water containing particles, although more research is needed to incorporate a range of particle sizes instead of an individual particle size.

2.5.1 Particle Fouling: Mass Transport of Colloidal Suspension

Fouling materials inside a membrane system experience various mass transport mechanisms and interactions such as adsorption and charge attraction, which determine the location of foulants and the extent of fouling. The adsorption and charge attraction might have more important roles on fouling in short range interactions. Bowen and Jenner (1995) argued that the particle-particle interaction model including electrostatic forces showed an excellent agreement with the predicted ultrafiltration rate in dead-end mode.

Mass transport mechanisms that carry potential foulants toward the membrane include convection (permeation drag) by applied transmembrane pressure and sedimentation by gravity. Several transport mechanisms carry materials away from the membrane (*i.e.*, back transport) in crossflow mode: Brownian diffusion, shear-induced diffusion, and inertial lift (Belfort, Davis, and Zydney 1994).

Brownian back diffusion is one of the major back transport mechanisms. The diffusion is produced by a concentration gradient, since high concentrations of rejected materials are created on or near the membrane surface due to the mass transport onto that surface. Other back transport mechanisms such as the shear-induced diffusion and the inertial lift originate from fluid dynamics in the crossflow mode. The magnitude of each mechanism in back transport is a function of particle size. The relative importance of each mechanism may be expressed as velocities. The velocities of particle transport by the Brownian diffusion, v_B, the shear induced diffusion, v_S, and the inertial lift, v_L, can be estimated as shown in Equations 2.15, 2.16, and 2.17 (Wiesner, Clark, and Mallevialle 1989; Wiesner and Aptel 1996):

$$v_B = kT/(3\pi\ \mu\ R\ d_P) \qquad (2.15)$$

$$v_L = u_0^2\ d_P^3\ \rho\ /\ (32\ \mu\ R^2) \qquad (2.16)$$

$$v_s = 0.05\ u_0\ d_P^2\ /\ (4\ R^2) \qquad (2.17)$$

where k= Boltzmann constant; T=absolute temperature; μ=viscosity of water; R= radius of a hollow fiber or capillary membrane; d_P= particle diameter; and u_o= centerline maximum velocity.

The Brownian diffusion is inversely proportional to particle diameter, whereas the shear-induced diffusion and the inertial lift are proportional to the second and third orders of particle diameter, respectively. Therefore, small particles are transported predominantly by Brownian diffusion. As particles become bigger, the velocity by Brownian diffusion diminishes, and the velocities by shear-induced diffusion and inertial lift become the main back transport mechanism in order.

When the back transport velocities are calculated under typical ultrafiltration conditions, a range of particles sizes between 0.1 μm and 1 μm have the minimal back transport velocity (Wiesner, Clark, and Mallevialle 1989; Lahoussine-Turcaud, Wiesner, and Bottero 1990; Lahoussine-Turcaud et al. 1990; Wiesner and Chellam 1992). If particles whose size is close to the nominal pore size of the membrane have the minimal back transport velocity under certain conditions, the particles could penetrate easily to block the pores and irreversibly foul the membrane.

Therefore, particle size distributions of feed water might have an important role in particle fouling. For instance, Lahoussine-Turcaud et al., (1990) revealed that coagulation shifted particles from colloidal size to bigger flocs and reduced the fouling rate, presumably by increasing back transport by shear-induced diffusion and inertial lift.

The studies on mass transport mechanisms of colloidal particles have increased the fundamental understanding of the effects of operational conditions such as cross-flow velocity and particle size (AWWA Membrane Technology Research Committee 1998). However, the mass transport mechanisms can explain only an individual particle trajectory. In reality, natural suspensions are quite heterodisperse, containing particles from less than 0.1 μm to 100 μm or more, and effects of such heterodispersity on cake porosity and filtrate flux are not yet fully understood. The particle size distribution in the deposited cake will be different from that in feed water, because a specific range of particles (e.g., 0.1 to 1 μm of particles) is deposited preferentially under typical conditions of cross-flow operation. Furthermore, the presence of particles will affect NOM removal and fouling, so the combined effects of NOM and particle fouling are even less understood and therefore require more research effort.

2.5.2 NOM Fouling

In addition to particulate matter, NOM has also shown a great tendency to foul membranes. Various solution characteristics such as pH, divalent ion concentrations, and ionic strength influence adsorption of NOM by physical and, especially, electrostatic interactions. In this section, their effects on NOM fouling are discussed in detail.

NOM Rejection and Adsorption

NOM rejection by MF or UF has been relatively low compared to NF or RO. In general, nanofiltration showed greater than 85% of NOM removal, whereas NOM removal varied from 5% to 40% in ultrafiltration (Mallevialle, Anselme, and Marsigny 1989; Chellam et al., 1997).

The variation in NOM rejection by UF depends primarily on the relationship between the NOM size distribution and membrane pore size. Laine et al., (1989) investigated effects of NOM size distribution on TOC removal with membranes of two nominal molecular weight cut-offs (MWCO), i.e., 5 kDa and 100 kDa. The apparent molecular weight distribution (AMW) of NOM in the source water was mostly smaller than 1 kDa or larger than 100 kDa; thus, TOC removals by the two membranes were similar. Since the AMW of NOM is different from source to source (Collins, Amy, and King 1985), selecting a membrane with an appropriate nominal pore size is important to obtain better NOM removal.

Although the rejection of NOM by MF or UF is relatively low, meaning that very small amounts of NOM are accumulated on the membrane surface, the reduction of water flux was substantial in several reported cases (Jonsson and Jonsson 1995, Chang 1996). According to Chang (1996), the SEM images after 9 hours of ultrafiltration of a surface water (Lake Pleasant, Seattle WA) showed that a NOM gel layer was formed on the membrane surface while the flux declined to 65% of the initial flux. The thickness of deposited organic matter was extremely small (only 0.5 μm), which implied that even an infinitesimal amount of NOM would foul membranes. Therefore, some NOM can be adsorbed and cause a hydraulic resistance on membranes, although most NOM has small enough molecular weights to pass through the pores of MF or UF.

NOM adsorption is known to be responsible for most of the irreversible fouling in membrane processes. The investigation of the fouled layer by SEM (Mallevialle, Anselme, and

Marsigny 1989) showed that NOM acted like a glue between a membrane and an inorganic cake layer. Although the deposited materials on the membrane from ultrafiltration of groundwater were mostly inorganics (i.e., 72% of ash by elemental analyses), organic carbon comprised 96% of the total carbon in the deposit, leading to the conclusion that organic carbon contributed greatly to fouling.

pH Effects

Solution chemistry has effects on shapes and sizes of NOM and on electrostatic repulsion between NOM and membranes. At most natural pH values, shapes of NOM are described as flexible and linear random coils because of the electrostatic repulsion between intramolecules in NOM. However, NOM becomes granular and compact in shape at low pH or high ionic strength, and in the presence of divalent cations. Therefore, the nominal sizes are reduced in those conditions because of less repulsive interactions between functional groups in NOM.

Furthermore, interactions between NOM and a membrane also become less repulsive due to fewer negative charges on NOM under these conditions. For instance, Hong and Elimelech (1997) showed that the rejection of NOM decreased at low pH. After an initial buildup of NOM on this surface leading to a higher diffusive gradient, the passage through the membrane was enhanced because of both the reduced size of NOM and the reduced repulsive interaction between NOM and the membrane. In addition, Jucker and Clark (1994) observed the increased adsorption of humic substances on the membrane surface at low pH (i.e., pH 4) due to the less negative repulsion and the increased diffusivity by the reduced sizes. However, note that little research has been performed previously with relatively high pH in UF, as would occur after softening.

Ionic Strength

High ionic strength causes similar effects as the low pH, i.e., increased NOM deposition because of the less repulsive interactions between NOM and membrane and a dense deposition layer because of the compact shapes of NOM. These phenomena occur due to the neutralization of charges by protons at low pH, whereas the double layer compression is the cause of changes in electrostatic interactions at the high ionic strength (Hong and Elimelech 1997).

However, contrary to the predicted severe fouling, some research indicated that small sizes of NOM could enhance the back transport by high diffusivity, and there is a possibility of coagulation of colloids by destabilization at high ionic strength (Wiesner and Chellam 1992; Bacchin, Aimar, and Sanchez 1996; Harmant and Aimar 1998).

Divalent Ions

The presence of divalent ions, especially calcium ions, has the same electrostatic effects on NOM shapes and interactions as the high ionic strength condition. Furthermore, calcium ions can form complexes with NOM. Jucker and Clark (1994) found that calcium was adsorbed in proportion to the amount of humic acid adsorbed, which implied that calcium was bound with humic acid. In addition, Hong and Elimelech (1997) showed that each calcium ion had a capability to complex with two carboxyl groups in humic acid. The authors implied that NOM became insoluble and precipitated with humic macromolecules at high Ca^{2+} concentration.

Furthermore, Yoon et al., (1998) found that NOM adsorption by the complexation of Ca^{2+} overcame the repulsive interaction by more negative charges from a high concentration of OH^-. Therefore, the flux decline was dramatically rapid in the presence of Ca^{2+} at high pH, which is the specific condition for softening. Although the results were obtained from nanofiltration, the results might have an implication on the use of UF in a softening plant, namely that pH might need to be adjusted prior to the membrane process.

Effects of Different NOM Fractions

In addition to solution chemistry, different components of NOM have specific effects on membrane fouling. Jucker and Clark (1994) found that humic acid (HA) had a greater adsorption capacity than fulvic acid (FA) to a relatively hydrophobic membrane. The hydrophobic properties and larger molecular weight (which rendered the great number of attachment sites) of HA increased the adsorption on the membrane. This result implies that pretreatment such as coagulation or softening, which can achieve significant removals of HA, are likely to benefit membrane processes.

In addition, NOM samples with different apparent molecular weight (AMW) showed different fouling behavior in UF (Lin, Lin, and Hao 2000). The authors divided the commercial humic acid (Aldrich Co.) into four fractions of AMW: 180-650 Da, 650-2200 Da, 2200-6500 Da, and 6500-22600 Da. The worst flux decline was found in the largest AMW (6500-22600 Da) range. And, removal of TOC in the largest AMW range was the greatest.

The effects of hydrophobic or hydrophilic components of NOM on membrane fouling are unclear, however. Nilson and DiGiano (1996) found that a hydrophobic fraction of a river water was responsible for most of the flux decline. The result was consistent with the idea that fouling increases with increasing molecular weight because the hydrophilic fraction was mostly smaller than 3000 Da, whereas the hydrophobic fraction had a higher concentration in the 10-30 kD range. However, Carroll et al., (2000) showed that filtrate rate for the neutral hydrophilic fraction of a river water declined by approximately 40% at the end of run, whereas the filtrate rate for both the strong and the weak hydrophobic fractions declined by only approximately 16%. Moreover, the commercial humic acid (Aldrich Co.) showed little difference in fouling from both the hydrophobic and the hydrophilic fractions (Lin, Lin, and Hao 2000). The discrepancies in fouling tendency by hydrophobic fractions are questionable when relative ratios between hydrophobic and hydrophilic components are different for different locations and sources (*i.e.*, natural or commercial). Table 2.4 shows that the commercial product (Aldrich Co.) contains generally a greater hydrophobic fraction. It is known that the extraction method to prepare the commercial humic substances tends to concentrate the more hydrophobic fraction of NOM. Results of membrane performance using the Aldrich humic acid, therefore, cannot be translated directly to real waters.

As mentioned in the section on NOM characteristics, some researchers have used the GC-pyrolysis-MS method to identify foulants from filtrate or deposited materials. Lahoussine-Turcaud et al., (1990) analyzed the NOM fraction of the Seine River water and the coagulated water by this method. The results indicated that coagulation removed polyhydroxyaromatic (PHA) materials to a greater extent than polysaccharide; thus, the relative fraction of polysaccharide to the DOC of the coagulated water was increased. The results implied that polysaccharide had a major role in membrane fouling after coagulation because of its high fraction in the feed water. In addition, the research by Mallevialle, Anselme, and Marsigny (1989) compared the relative fraction of each type of compound in feed water, filtrate, and concentrate in UF. Carbohydrate (i.e., mono- and polysaccharide) was primarily in the deposit or concentrate water. Proteins and PHA were removed from feed water but were not in filtrate, which means that they were adsorbed onto the membrane. Amino-sugar and low molecular weight materials preferentially crossed the membrane and reached the filtrate. Further, the analyses on deposited organics suggested that the concentration of carbohydrate was high in the deposit although it has low adsorption affinity. Also, protein was enriched in the deposit, which implied that protein might be the most responsible foulant.

Mackey and Wiesner (1999) investigated membrane fouling with model compounds for each fraction, i.e., polygalacturonic acid for polysaccharide, rosolic acid for PHA, and bovine serum albumin for protein. The results showed that protein was apparently the main foulant. However, polysaccharide showed bigger effects on fouling based on the fouling rate per mass. The authors indicated that the results might be different if a polysaccharide with a slightly different molecular weight distribution was used.

Although considerable research has been devoted to understanding NOM fouling on membranes, more research is required to fully understand how different characteristics of NOM contribute to fouling.

Table 2.4
Relative fraction of hydrophobic and hydrophilic NOM from various sources

Sources	Hydrophobic	Hydrophilic	Reference
Humic substances (Aldrich Co.)	85%	15%	Lin, Lin, and Hao 2000
Peat fulvic acid*	73%	27%	
Grasse River, NY	57%	43%	
Floridan Aquifer, FL	58%	42%	Collins, Amy, and King 1985
Cobble Mtn reservoir, MA	49%	51%	
Colorado River+, NV	35%	65%	
Tar River, NC	50%	50%	Nilson and DiGiano 1996
Moorabool River, Australia	52%	48%	Carroll *et al.* 2000

*: extracted from Michigan peat soil,
+: known as a very low humic acid water (Hwang et. al 1999)

2.5.3 Surface Characteristics of Fouled Membranes

Surface chemistry, or phenomena that occur on membrane surfaces, often determine the extent and pattern of membrane fouling. To investigate effects of surface chemistry on membrane fouling, three important surface characteristics of fouled membranes are presented below: (1) zeta potential, (2) morphology, and (3) elemental composition analysis.

Zeta Potential

A polymeric membrane surface generates certain charges and leads to an electric potential near the surface when it is contacted with an electrolyte solution. The membrane surface and the surrounding liquid layer can then be described by the electrical double layer model.

Three of the most popular polymer materials for manufacturing membranes are cellulose acetate, polyethersulfone, and polysulfone. Each polymeric membrane contains either ion groups ($-SO_3^-$) or ionizable functional groups ($-NH_2$ and $-COOH$). These groups exhibit certain charges under specific conditions of the solution.

The surface charge of the double layer, often measured by the zeta potential, determines the electrokinetic aspects of mass transfer across a membrane as well as influences the interactions between membranes and foulants (Childress and Elimelech 1996, Pontie et al., 1997, Cho 1998). Electrostatic interactions on pore surfaces have been hypothesized to be more responsible for irreversible fouling than other physical interactions in MF and UF (Bowen and Jenner 1995).

The zeta potential of a membrane surface is affected by several factors such as pH, ionic strength, and the presence of divalent ions and humic substances in solution. In general, hydrogen ion concentration plays a dominant role in determining the zeta potential of a surface. The surface charge tends to be positive at low pH and negative at high pH. The points of zero charge of most of membranes are usually between pH 3 and pH 5.5; therefore membrane surfaces have negative charges in natural water. Hong and Elimelech (1997) showed that negative charges on the membrane surface helped to reduce deposition of foulants on the membrane, thus leading to a higher filtrate rate (less fouling). This occurred because of the repulsive forces between the membrane and the colloidal or organic substances, which also had negative charges in natural water.

The zeta potential also showed that fouling could be different depending on membrane sizes. As the amount of humic substances in the feed water was increased, Jucker and Clark (1994) measured the zeta potential of two ultrafilters: PM30 (polysulfone, 30 kDa, Amicon Co.) and XM50 (poly acrylonitrile-co-vinyl chloride, 50 kDa, Amicon Co.). They found that adsorption of both HA and FA on the membranes increased as pH decreased from 7 to 3.2. As pH decreased, the charges on both humic substances and the membrane became less negative, leading to lower repulsive forces between the humic substances and the negatively charged membrane surface, and thereby enhancing the adsorption of humic substances.

The increased adsorption of humic substances increased zeta potential in MF and UF, but decreased it in NF and RO (Childress and Elemelech 1996). To explain these differences, the authors postulated that the adsorption of humic substances must have occurred mostly inside the pore surface for MF and UF, but on the filtering surface for NF and RO. Note that waters that require softening or which have been softened generally have relatively high pH, and therefore might be expected to cause less fouling, at least by adsorption of humic substances.

Membrane Surface Morphology – SEM/AFM

As electron microscopy technologies have been developed, skin layer morphologies of polymeric membranes have been directly examined to understand properties of membranes such as pore structure, pore size distribution, porosity, and shape. The properties that are related to surface morphology are important factors in determining filtration rate and fouling of membranes.

Among electron microscopes, the transmission electron microscopy (TEM) and the scanning electron microscopy (SEM) are frequently used to investigate membrane surfaces (Kim et al., 1990, Kim et al., 1992, Kim and Fane 1994, Cho 1998). Generally, the TEM has a higher resolution (0.3~0.5 nm) than the SEM (0.7 nm). However, due to the method used to release electrons to samples, the TEM can examine only very fine structures, and so is not proper for three-dimensional images. Therefore, the SEM is a more popular method to take images of relatively rough polymeric membrane surfaces. Furthermore, the field emission SEM (FESEM) has been developed to reduce damages from the electron beam by using lower beam energy than the standard SEM. In addition, the recent development of atomic force microscopy (AFM) relieved the necessity of a drying procedure, which might alter pore shapes or deposited cakes of foulants (Mackey and Wiesner 1999).

Using the SEM and the TEM, Kim et al., (1990) revealed that structures of membrane surfaces significantly differed from membrane to membrane although they have the same nominal pore sizes. For instance, XM 100A (poly co-acrylonitrile-vinyl chloride, Amicon Co.) and MPS (polysulfone, Memtec Co.) had the same nominal molecular weight cut-off (*i.e.*, 100 kDa), but their SEM images made clear that XM 100A had more uniformly distributed pores and a relative rough pore structure, whereas MPS had irregularly spread pores and a very smooth pore structure. The pore size of membranes has usually been reported with a single value such as 0.01 μm or 10 kDa for ultrafiltration, but in fact membranes have a pore size distribution. Therefore, some particles or NOM that are smaller than the nominal pore size could be caught on the membrane, and some that are larger than the nominal pore size could pass through.

In addition, comparison of images of clean and fouled membrane provides qualitative proof of deposition and adsorption of foulants. The SEM or AFM images of a membrane surface are different for clean and fouled membranes. Mackey and Wiesner (1999) used three different NOM components (*i.e.*, protein, polyhydroxyaromatics, and polysaccharides) to create fouled membrane surfaces. The AFM images of the surface fouled with each component showed very different structures of fouling. Since the components were dissolved chemicals, the changes from the clean membrane were due to the adsorption of each component. Using the FESEM, Kim et al., (1992) illustrated that fouling by a high concentration solution of a protein (i.e., bovine serum albumin) occurred on the ultrafilter surface, not inside the pores.

Although information from the SEM/AFM is precise and direct about membrane surfaces, it is not quantitative; therefore, to understand the extent of fouling, it is necessary to investigate quantitative data that reflects fouling such as hydraulic filtration rate.

Elemental Composition Analysis – XPS

It has been widely recognized that the nature of the membrane surface is critical to performance. The quantitative elemental compositions of the topmost layer of membrane surfaces can be measured using X-ray photoelectron spectroscopy (XPS). Although advantages of XPS have been extensively proven in polymer applications, little research has used XPS analysis on membrane processes. Nevertheless, XPS appears to be a promising technique to aid understanding of membrane characteristics and performance.

The main features of XPS are a vacuum chamber, an X-ray gun, a sample manipulator, and an electron-energy analyzer. When a sample is irradiated with soft X-rays, photoelectrons are emitted from either the valence or core level. An electron from a higher level then fills the vacancy created from emitted photoelectrons. The energy released in filling the core vacancy results in the emission of an X-ray, which is collected by a lens system and focused into an energy analyzer. Detailed descriptions of XPS are found elsewhere (Briggs 1983, Munro and Singh 1993).

The photoelectrons escaping into the vacuum are counted as a function of their kinetic energy. The conservation of energy law allows the binding energy (E_{bin}) of the emitted photoelectrons to be calculated, where hν is the X-ray energy, ϕ is the sample work function, and E_{kin} is the kinetic energy of the photoelectron. The binding energies of core electrons are characteristics of individual elements and hence the elements are identifiable.

$$E_{bind} = h\nu - E_{kin} - \phi \qquad (2.18)$$

Oldani and Shock (1989) performed an elemental analysis of clean membranes by XPS. The results showed that XPS accurately measured the chemical composition of the 2-nm thick layer of the membrane, since it was consistent with those calculated from the chemical formulas of the membrane and also with infrared spectroscopy.

Jucker and Clark (1994) used XPS to evaluate Ca^{2+} binding on a membrane surface along with adsorption of humic substances. The authors observed that there were two different binding energies for Ca^{2+}: the characteristic of calcium phosphate compounds (347.2 ± 0.1 eV), which is the binding energy of Ca^{2+} in most cases and a unique value measured only in this research (349.2 –349.7 eV) when humic acid was adsorbed in a polysulfone membrane. The changes in the binding energy implied that calcium interacted with humic acid and altered its chemical nature in some manner.

The XPS of important inorganic compounds and organic matter in feed waters such as C, O, S, Ca^{2+}, Mg^{2+}, and Si^{2+} can be analyzed to evaluate changes in elemental composition by fouling. Results from elemental composition studies can illustrate the main fouling component.

2.6 PRETREATMENT BEFORE MEMBRANE PROCESSES

Although considerable research has been devoted to understanding fouling in membrane processes, rather less attention has been paid to determine effects of pretreatment on fouling. The purpose of pretreatment is to remove fouling materials (mainly particles and natural organic matter) before they can deteriorate the membrane performance.

NOM has been recognized as a main foulant in membrane processes, but the degree of fouling by either NOM or particles depends on how fouling occurs in specific conditions

(Champlin 2000). Carroll et al., (2000) showed that particle fouling was dominant in MF if waters were not pretreated, but NOM fouling became dominant when coagulation or PAC adsorption was used to pretreat the feed water. The results by Chellam et al., (1997) also implied that particles determined the properties of fouling during nanofiltration with conventional pretreatment. Of three pretreatment options considered (conventional treatment, MF, or UF), NOM removal was much greater (approximately 50%) with conventional treatment than with MF or UF (approximately 5%). However, the fouling was the greatest with conventional treatment. Based on the silt density index, which is a parameter indicative of particle fouling, particle fouling was the dominant cause of fouling (ASTM, 1995). Therefore, both NOM and particle fouling need to be examined to better understand fouling in ultrafiltration.

Numerous factors such as cost, available area, fouling rate, and quality of a finished water could affect the question as to what pretreatment should be chosen for a specific situation (Chellam, Serra, and Wiesner 1998, Kelley and Olson 1999). In addition, characteristics of feed water have a great role in determining the optimal pretreatment of membrane processes in drinking water systems.

2.6.1 Pretreatment Alternatives

Limited studies on pretreatment schemes for ultrafiltration have included coagulation (Wiesner, Clark, and Mallevialle 1989; Lahoussine-Turcaud, Wiesner, and Bottero 1990; Lahoussine-Turcaud et al., 1990; Laine, Clark, and Mallevialle 1990), and powdered or granular activated carbon (PAC/GAC) adsorption (Adham et al., 1991, Jancangelo et al., 1995, and Campos et al., 2000). Investigation of pretreatments for NF includes conventional treatment (Wilkes et al., 1996 and Chellam et al. 1997) and the use of membrane processes with larger sizes of pores (Chellam et al., 1997). In addition, dissolved air flotation (Braghetta et al., 1997) and coagulation (Carroll et al., 2000) were applied as a pretreatment to MF.

Table 2.5 summarizes the previous studies on pretreatment before membrane processes. For the combination of pretreatment and the membrane process, TOC removal is often significant. For instance, Jacangelo *et al.* (1995) observed that 82% TOC removal from a river water (the Mokelumn River, CA) was achieved with 90 mg/L of PAC addition prior to UF treatment, compared to 10-15% TOC removal by UF alone.

Coagulation

Coagulation in drinking water systems has been used as a primary process to remove particulate matter, including microorganisms, for many years. In the last 20 years, studies have shown that coagulation processes substantially remove NOM and DBP precursors. Therefore, coagulation has been investigated as a pretreatment of membrane processes because of its capability to remove the main foulants, i.e., NOM and particles, as well as the ability to use the existing infrastructure.

Table 2.5
Pretreatment Alternatives in Membrane Process

Reference	Pretreatment	NOM Removal by pretreatment	Membrane	Total NOM Removal
Lahoussine-Turcaud, Wiesner, and Bottero 1990	Coag.	40% UV 70-90% UV	UF (PS, 1kD)	NA
Lahoussine-Turcaud et al., 1990	Coag.	>85% UV	UF (PS, 1kD)	NA
Braghetta et al., 2000	In-line coag./settling	35% TOC	UF (NA)	35% TOC*
Jacangelo et al., 1995	PAC	NA (10-15% TOC only by UF)	UF (CD, 100kD)	82% TOC w/90mg/L PAC 38% TOC w/200mg/L PAC 23% TOC w/ 30mg/L PAC
Laine, Clark, and Mallevialle 1990	No Coag. PAC PAC-coag.	NA	UF (acrylic copolymer, 100kD)	42% NPOC 63% NPOC 84% NPOC 88% NPOC
	No Coag. PAC PAC-coag.	NA	UF (RC, 100kD)	43% NPOC 53% NPOC 85% NPOC 88% NPOC
Chellam et al., 1997	MF (PP 0.2 μm) UF (CE, 10kD) Conven.	5% TOC 5% TOC 50% TOC	NF (PES, 200D)	90-95% TOC
Wilkes et al., 1996	Conven.	40 % TOC	NF (Modified PA)	87% TOC
Braghetta et al., 1997	DAF	35-50% TOC w/45mg/L Alum	MF (PP, 0.2μm)	45-60% TOC
Carroll et al., 2000	Coag. PAC	46% TOC 68% TOC	MF (PP 0.2 μm)	NA
Freeman et al., 2000	No pretreat In-line coag. w/ alum	NA	UF (Proprietary PVDF)	3-10% 48-52%

PA: polyamide, PAN: poly acrilonitrile, PS: polysulfone, CD: celluloic derivatives, CE: celluloic ester, PP: polypropylene, PES: polyethersulfone, RC: regenerated cellulose, Coag.:coagulation, Conv.:conventional treatment, NA: not available *: No additional TOC removal by ultrafiltration

Coagulation impacts membrane fouling primarily by two mechanisms: reduction of fouling by changing small particles to bigger aggregates and adsorption of NOM by the precipitates formed in "sweep floc" coagulation. By adding coagulants, colloidal particles are destabilized and attached together, and then aggregated to bigger flocs, which can be removed by filtration or in the clarification process. During the destabilization by alum or iron, dissolved materials are adsorbed onto the $Al(OH)_3$ or $Fe(OH)_3$ aggregate or precipitated as an aluminum or iron-organic (e.g., aluminum fulvate), and removed from water.

The bigger particles formed by coagulation enhance the mass transport away from the membrane, i.e., the back transport mechanisms in crossflow operation. The greater back transport velocities reduce accumulation of foulants in the deposited layer. Wiesner, Clark, and Mallevialle (1989) measured the approximate average particle size while the pH and dose of coagulant were changed. The authors found that the biggest particle was formed and the least cake resistance was built when the zero zeta potential of particles was obtained.

Furthermore, there was a certain particle size that had the minimal predicted velocity of back transport, i.e., approximately a few micrometers under the typical operational conditions in ultrafiltration. With coagulation treatment, small particles with size similar to pore size were shifted to big particles that had significant back transport velocities. Therefore, the increase in particle sizes changed the fouling characteristics from the pore blockage (irreversible fouling) to the cake layer formation (reversible fouling).

The formation of flocs in coagulation showed other positive effects on the extent as well as the rate of fouling. A study by Braghetta et al., (2000) showed that use of chemical coagulants prior to ultrafiltration achieved in the range of 25 to 50% higher flux than that demonstrated for the untreated raw waters. In addition, coagulation pretreatment improved the fouling layer properties, i.e., it reduced irreversible fouling (Lahoussine-Turcaud, Wiesner, and Bottero 1990; Lahoussine-Turcaud et al., 1990). Although the flux decline of pretreated water was similar to that of raw water, the flux recovery by surface wash was greater with the coagulation pretreatment. The results implied that aggregating fine particles increased fouling cake permeability or conditioned the cake by incorporating fine particles into highly porous flocs and adsorbing dissolved materials into flocs (Wiesner and Aptel 1996, Carroll et al., 2000)

The removal of NOM, i.e., foulants in ultrafiltration, was also substantial in coagulation. The reported removals of NOM by conventional treatment including coagulation and softening in full-scale plants varied from 25% to 60% depending on applied processes, types of coagulant used, and water sources (Kavanaugh 1978; Collins, Amy, and Steelink 1986; Lykin and Clark 1989; Vik and Eikebrokk 1989). Since NOM removal has been recognized to be important to control DBPs in conventional treatment, substantial research has been conducted to provide comprehensive assessments of NOM removals in the processes (Randtke et al., 1994, White et al., 1997).

In general, coagulation removes primarily high molecular weight fractions (usually bigger than 5 kDa) and hydrophobic fractions of NOM; these fractions are more reactive and produce more DBPs than hydrophilic fractions. Humic acid is a major component of the high molecular weight fraction in natural water. The chlorination of humic acid yielded more than twice as much chloroform formation as fulvic acid, which has smaller molecular weight. Therefore, coagulation was recognized as an effective process to reduce THM precursors in drinking water systems (Collins, Amy, and King 1985; Vik and Eikebrokk 1989; Chow et al., 1999).

Collins, Amy, and King (1985) mentioned that all treatment including coagulation and softening were more effective in removing the higher molecular weight fraction (5,000-

10,000 Da) and less effective in removing the lower molecular weight fraction (<500 Da). Also, the authors characterized NOM by different acidic groups; fulvic acid generally had a higher charge density than corresponding humic acid. And the carboxylic acidic content (the main functional group determining charge density of NOM) appears to be inversely related to molecular weight. Therefore, they suggested that NOM with the highest content of acidic functional groups would be the most difficult to be destabilized by coagulation.

In addition, the results from fractionation of a surface water by Carroll et al., (2000) were consistent with the fact that hydrophobic NOM was preferably removed in alum coagulation. The fractions of the surface water were strongly hydrophobic, weakly hydrophobic, charged hydrophilic, and neutral hydrophilic fractions. The fractionation revealed that the hydrophobic or charged fractions of NOM were removed in substantial amounts, but the neutral hydrophilic fraction was not removed at all.

PAC Adsorption

Although coagulation can achieve considerable removal of NOM, Laine, Clark, and Mallevialle (1990) argued that PAC adsorption was a better option compared to coagulation as a pretreatment of UF. In addition to better adsorption of NOM by PAC, the authors mentioned that the coagulation process removed the high molecular weight fraction of NOM, which would be also easily removed by UF. On the other hand, PAC adsorption removed more of the low molecular weight compounds and not as much of the high molecular weight compounds.

PAC adsorption transfers dissolved organic matter onto a particulate surface (i.e., the powdered activated carbon) which can be easily removed by filtration; thus PAC pretreatment helps to reduce NOM fouling. The pilot data from Braghetta et al., (2000) showed increases in the specific flux with additions of 10 to 20 mg/L of PAC to the raw water. The authors explained that the improvement of the specific flux was attributed in part to the adsorption of small organics, and also to possible surface scouring by recirculation of PAC across the membrane surface. Also, Jacangelo et al., (1995) showed that the addition of 20 mg/L of PAC retarded the fouling without any buildup of resistance by PAC although the dose of PAC was not enough to completely remove the NOM in the feed water.

However, some research indicated that PAC addition showed little increases in water flux although NOM was substantially removed (Laine, Clark, and Mallevialle 1990; Carroll et al., 2000). Also, Wiesner and Chellam (1992) warned that smaller sizes of PAC, i.e., smaller than approximately 5 μm, could foul the membrane because the accumulated PAC can add a resistance of the cake layer. Therefore, careful considerations are necessary to apply PAC to a UF system, although the small sizes of PAC increase the adsorption kinetics and so reduce the required contact time.

All pretreatment alternatives presented above achieved substantial removal of NOM, but showed little increase in water flux, which is the practical indicative parameter in membrane process. Therefore, further research on fouling mechanisms by either NOM, which is difficult to remove by all pretreatment alternatives, or particles, which are stabilized with the very low concentration of NOM in solution, is necessary.

2.6.2 Softening Pretreatment

As shown in Table 2.4, studies on pretreatment using conventional treatment have tended to focus on coagulation rather than on softening. However, many utilities in Texas and throughout the central U.S. use precipitative softening, and most of those plants require modifications to meet the new and expected regulations. Therefore, research designed to elucidate effects of softening pretreatment on membrane performance is necessary.

2.7 SUMMARY

Ultrafiltration is a promising process to obtain a high quality drinking water with respect to microorganisms, and so secure the public health with respect to water-borne diseases. However, fouling by NOM and particulate matter is a major concern in membrane processes. Softening or coagulation can be a feasible pretreatment for UF because of the removal of some NOM and particulate prior to ultrafiltration and the capability to use the existing infrastructure.

One of the main problems in the use of membranes is particle fouling, which is the accumulation of particles on the membrane surface; particle fouling can be predicted for the crossflow mode based on the mass transport mechanisms in UF. Particles with the sizes having the minimal back transport velocity can be deposited easily on a membrane surface under the typical operational condition in UF. However, the mass transport mechanism explains only discrete particles. Much research, therefore, needs to be done to understand interactions between particles of different sizes and also between particles and NOM.

In addition to particles, NOM is often the most dominant foulant. Therefore, effects of NOM characteristics such as hydrophobicity and molecular weight on the extent of fouling as well as on the removal in membrane processes, have been extensively investigated. In general, hydrophobic NOM showed greater adsorption and higher fouling tendency than hydrophilic NOM in ultrafiltration. Also, several processes such as coagulation and PAC adsorption were evaluated to assess its feasibility as a pretreatment option for membrane processes. However, little research has investigated effects of enhanced softening, i.e., the extents of NOM removal in softening, on NOM fouling in UF. This research was designed to examine effects of softening performance for particle and NOM on fouling in UF.

CHAPTER 3
EXPERIMENTAL METHODS

3.1 OVERVIEW

The experimental protocol for this work was designed to simulate softening followed by an ultrafiltration process to investigate effects of softening pretreatment on fouling in ultrafiltration. A schematic diagram for the experimental protocol is shown in Figure 3.1.

Several source waters were introduced to understand effects of the removal of individual constituents in softening on ultrafiltration fouling. The source waters were synthetic waters with inorganic constituents only, synthetic organic waters, and two natural waters. At first, the natural waters were characterized in terms of inorganic (especially hardness ions and alkalinity) and organic constituents. Based on the results from the characterization of the natural waters, the synthetic waters were made to simulate those natural waters.

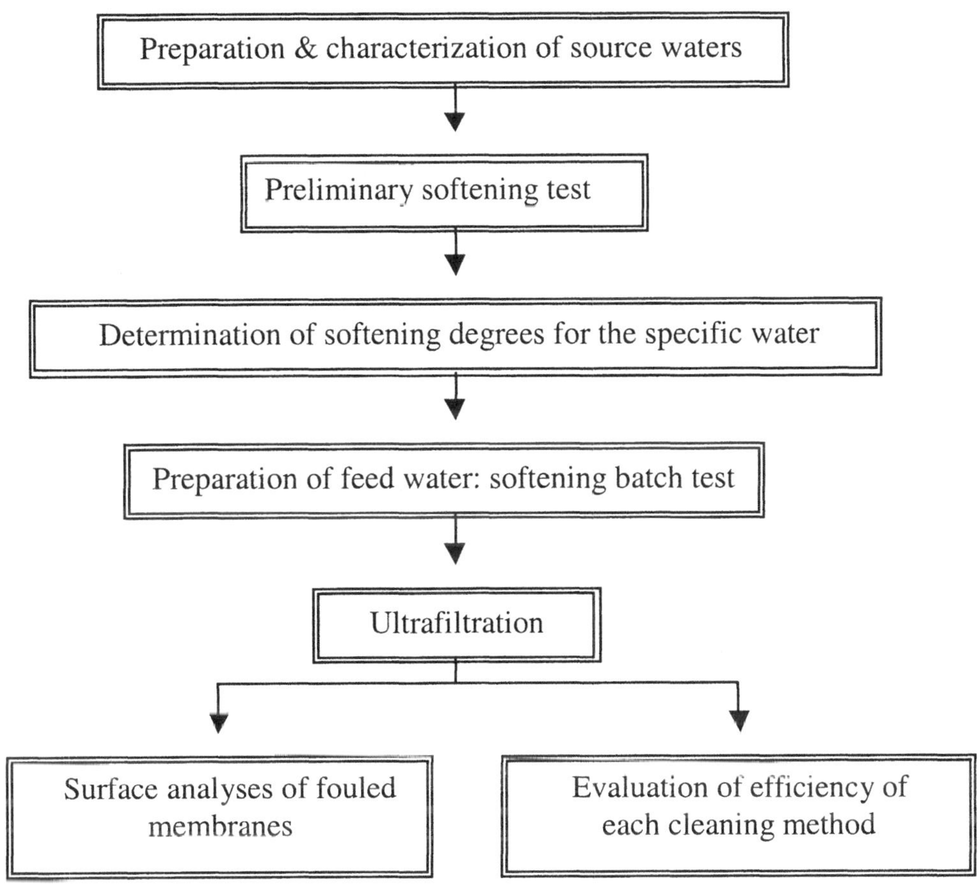

Figure 3.1 Flow sheet of a general experimental plan

The waters were then softened to different extents. Based on the softening performance, two or three levels of softening were selected to further investigate effects of the extent of softening on membrane fouling. The three possible levels of softening are referred to herein as standard softening, enhanced softening, and Mg softening conditions. The "standard softening" condition was chosen on the basis of softening parameters alone (i.e., the minimum lime dose to achieve excellent calcium removals). The second "enhanced softening" condition represented the maximum lime dose without forming voluminous precipitates of $Mg(OH)_2$, and the third "Mg softening" condition was a higher lime dose where the precipitation of $Mg(OH)_2$ would occur to a significant extent and remove a high percentage of the magnesium.

The waters were softened to the specific levels noted above and, after solid-liquid separation, were introduced to ultrafiltration to investigate effects of softening pretreatment on the process. The softening performance was evaluated based on several key water parameters and effects on fouling.

The ultrafiltration performance was monitored based on flux decline behavior. Then, some fouled membranes were further investigated with scanning electron microscopy (SEM) or X-ray photoelectron spectroscopy (XPS) to learn details of the physics and chemistry of fouling. Several membranes were evaluated to determine the efficiency of three cleaning methods: a surface (distilled water) wash for general deposition on the membrane surface, a caustic wash for organic components, and an acidic wash for inorganic foulants. The efficiency of these cleaning methods was evaluated in terms of the recovery of the clean water specific flux after each cleaning method.

3.2 MATERIALS

3.2.1 Membranes

Several ultrafilter materials were considered in terms of hydrophilic properties, which have shown dramatic effects on membrane fouling. Two materials with great differences in hydrophilicity were selected for this research: polysulfone for a hydrophobic membrane and regenerated cellulose for a hydrophilic membrane. Both membranes were purchased from Millipore Company (Bedford, MA). The membranes were flat sheets with 30 cm^2 of filter area since the ultrafiltration system was a plate and frame module, which was also purchased from Millipore Company. Their nominal molecular weight cut-offs (provided by manufacturer) were 10 kDa. The clean water flux characteristics of these membranes measured in this research are listed in Table 3.1. The clean water flux varied between different sheets of each type of membrane, *i.e.*, polysulfone had a coefficient of variation (standard deviation/mean) of 29% while regenerated cellulose had 34%. The clean water specific flux of the two membrane types showed a lower coefficient of variation than the clean water flux (*i.e.*, 18% for polysulfone and 31% for regenerated cellulose) since the flux was normalized with the transmembrane pressure, which could vary during different operations.

Table 3.1
Physical characteristics of the UF membranes

Membrane Type	MWCO (Da)	Clean water flux*, J_o (L/m^2-hr at 20 °C)	Clean water specific flux*, J_{so} (L/m^2-hr-kPa at 20 °C)
Polysulfone (PS)	10,000	150 ± 43	1.65 ± 0.46
Regenerated cellulose (RC)	10,000	47 ± 16	0.55 ± 0.12

* Mean and standard deviation

3.2.2 Natural Waters

Two natural water sources were selected and characterized for this research: the water supplies for the City of Austin (Lake Austin) and St. Louis County Water (the Missouri River). Raw water from Lake Austin was collected in a large volume (approximately 500L) at one time in November 1998. The water was collected from the header pipe following the raw water intake pumps at the Ullrich Water Treatment Plant, a plant that currently practices softening. Raw water from the Missouri River was collected at one time from the Howard Bend Treatment Facility, which serves the city of St. Louis and St. Louis and St. Charles counties. The water was shipped from St. Louis to Austin in September 1999. The raw waters have been stored in a 4°C refrigerator since collection. The raw water characteristics, including pH, Ca^{2+}, Mg^{2+}, and DOC, were monitored at times and showed little changes in water quality over many months.

The collection of raw water in a large volume was intended to allow all tests to be conducted on a single batch, thus avoiding source water variability, which causes complications in data interpretation. Since the Missouri River water has high turbidity (approximately 300 NTU), however, the turbidity for the Missouri River water could be changed during the long storage time due to adhesion of particles to the walls; therefore, experiments on this water were performed in a short period and the raw water quality was measured at the beginning of each experiment.

To characterize the natural water, several key water quality parameters were measured in the laboratory, including (1) pH, (2) hardness ions such as calcium and magnesium, (3) alkalinity, (4) organic carbon including total organic carbon (TOC), dissolved organic carbon (DOC), and hydrophobic DOC, and (5) the ultraviolet absorbance at 254nm (UV_{254}). In addition, the specific UV absorbance (SUVA) was calculated as the ratio of UV_{254} to DOC. The analytical methods used for raw water characterization are described in Section 3.5 and Ralls (1999).

3.2.3 Synthetic Inorganic Water

Synthetic water with only inorganic constituents was produced to investigate inorganic fouling exclusively, *i.e.*, without NOM effects on both softening and membrane performance. This water mimics the inorganic constituents of Lake Austin water and contains only Ca^{2+}, Mg^{2+}, Na^+, Cl^-, and CO_3^{2-}. The calculation of concentrations of each constituent is shown in Table 3.2. The concentrations of hardness ions were calculated from the raw water concentrations, i.e., 154 mg/L as $CaCO_3$ of calcium and 80 mg/L as $CaCO_3$ of magnesium. Chloride was added with

$CaCl_2$ and $MgCl_2$. Thus, Na^+ was introduced as $NaHCO_3$ to adjust alkalinity to Lake Austin water, i.e., 169 mg/L as $CaCO_3$.

The synthetic water was used for various scenarios of softening pretreatment including lime softening only, lime softening with pH adjustment (re-carbonation), or lime soda softening. Then, the scenario of softening with the least inorganic fouling on the membrane surface was considered for subsequent experiments to investigate the effects of NOM and particles on membrane performance. In addition, the synthetic water was used to investigate inorganic fouling on the membranes after different levels of softening were applied as pretreatment before ultrafiltration.

3.2.4 Simple Surrogates for Natural Organic Matter

Simple organic components were considered as surrogates for NOM. Since NOM has various functional groups and complicated structures, it is difficult to understand the role of particular functional groups or the substructure of NOM in water treatment. Previous research on NOM components in membrane processes indicated that polysaccharides were suspected to be the main foulants for membrane processes (Mallevialle, Anselme, and Marsigny 1989; Lahoussine-Turcaud et al., 1990; Mackey 2000). An increase (from 36% to 47%) in the relative contribution of polysaccharides to the DOC was noticed after coagulation treatment (Lahoussine-Turcaud et al., 1990). Polysaccharides were also found to form an important part of dissolved macromolecular solutes present in the fouling cake (Mallevialle, Anselme, and Marsigny 1989; Mackey 2000). Because they are a major NOM component and have been found to be the largest cause of fouling, further investigation of polysaccharides was necessary to understand fouling in ultrafiltration.

Table 3.2
Example of calculation for each constituent concentration in the synthetic inorganic water

Constituent	Calculation
Ca^{2+}	154 mg/L as $CaCO_3$ = 3.08 meq/L = 1.54 mmol/L = 170.9 mg/L as $CaCl_2$
Mg^{2+}	80 mg/L as $CaCO_3$ = 1.6 meq/L = 0.8 mmol/L = 76.2 mg/L as $MgCl_2$
Cl^-	2×1.54 mmol $+ 2 \times 0.78$ mmol $= 3.08 + 1.6 = 4.68$ mmol/L
Na^+	Alkalinity -($[Ca^{2+}]\times2+[Mg^{2+}]\times2-[Cl]$)

Table 3.3
Characteristics properties of some NOM and polysaccharides

Compounds	Molar mass (kDa)	Charge	Poly-dispersivity*	Comments	Ref.
Polyacrylic acid	37-1360	Negative, COO^-	1.6	Fairly monodisperse, linear flexible polyelectrolytes	Wilkison et al., (1999)
Alginic acid	48-2000	Negative, COO^-	2.6	Polydisperse, semi-flexible, negatively charged polyelectrolytes; extended random coils	Buffle et al., (1998)
Dextran	2-3000	Neutral	-	Neutral; dense coil	Buffle et al., (1998)
UKGSAHA**	20	Negative, mainly $COOH^-$	Large	Very polydisperse, semi-flexible, branched, negatively charged	Wilkison et al., (1999)
SR NOM#	Unknown	Negative, mainly COO^-	Large	Very polydisperse, negatively charged	Wilkison et al., (1999)

* *Polydispersivity is reflected by the polydispersivity index (PI)= MW/MN, Mw: weight-average molecular weight; MN: number-average molecular weight.*
** *UKGSAHA represents U.K. Geological Survey Aquatic Humic Acid.*
\# *SR NOM represents Suwannee River NOM*

In preliminary tests, several polysaccharides including dextran, alginic acid, and polygalacturonic acid were evaluated for their removals in softening. In addition, a smaller carbohydrate, tannic acid, was investigated. Dextran is a relatively dense and neutral polysaccharide, whereas alginic acid and polygalacturonic acid are negatively charged in the pH range of softening because of the carboxylic acid functional groups. Tannic acid is not a polysaccharide but rather an oligosaccharide with free phenolic groups. The phenolic groups are expected to be de-protonated at the pH of softening, and hence might adsorb to surfaces to a great extent.

The screening tests revealed that the NOM surrogates could be dextran and alginic acid, since their behavior in the softening process was more analogous to Lake Austin water than the polygalacturonic acid and tannic acid solutions; these results are shown in detail in Chapter 4. The results from polygalacturonic acid showed that the DOC removal was almost complete at the standard softening condition (i.e., 125 mg/L lime dose), and no further DOC removal occurred with increasing lime doses. Therefore, the polygalacturonic acid was not useful as a surrogate for NOM in this research because the degree of softening made no difference in the removal. In addition, tannic acid was tested with the same procedure as the other organics. However, in trying to make a 4 mg/L (as carbon) solution in the synthetic hard water from a concentrated tannic acid solution (in distilled deionized water), the solution became quite turbid, indicating that a precipitate was formed. After filtration through a 0.45 μm membrane, the DOC was measured as 0.8 mg/L C; the fact that this measure was far less than the 4 mg/L target confirmed the precipitation. Hence, tannic acid apparently forms a precipitate with the calcium or magnesium in a hard water at a pH of approximately 8.3, and is therefore an inappropriate surrogate for NOM in this research.Characteristics of the selected polysaccharides (i.e., alginic acid and dextran) are compared with polyacrylic acid and some NOM in Table 3.3. The molecular weights range widely from 2 to 3000 kDa. The polydispersivity shows that NOM is much more polydisperse than a polysaccharide. Polydispersivity is one of the major characteristics of NOM and creates complicated reactions in water treatment.

Structural images of alginic acid and dextran are shown in Figure 3.2. Alginic acid has many carboxylic acid functional groups (i.e., COOH) and consists mainly of D-mannuronic acid and L-glucuronic acid. The structure of dextran shows no ionizable functional groups, leading to neutral molecules.

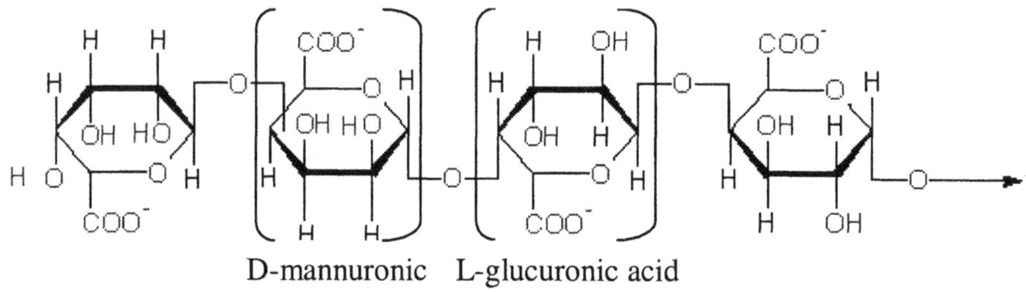

(a) alginic acid

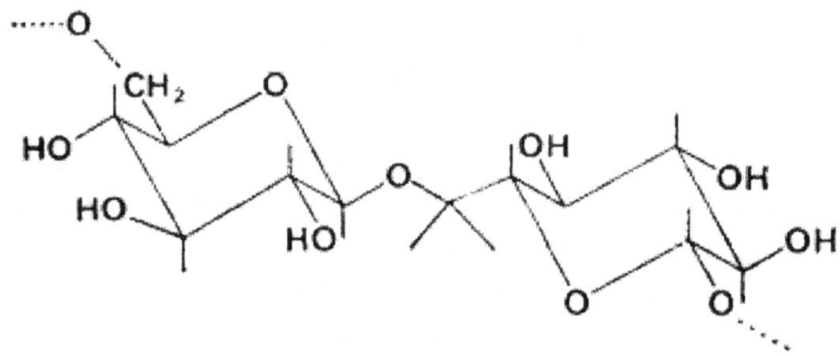

(b) dextran

Figure 3.2 Structural images of (a) alginic acid and (b) dextran

3.3 SOFTENING

3.3.1 Softening Jar Test

Softening jar tests for the raw water characterization and for preliminary tests of synthetic waters were conducted using a six-place gang-stirrer jar tester. The jars are square (11.5 cm inside) with floating covers to prevent carbon dioxide from being dissolved into the water during tests, thus affecting pH and impacting precipitation of Ca^{2+} and Mg^{2+}. Details of the design of the jars are available in Sun (1998) and Ralls (1999).

Jar tests were performed on one liter of water. The water was collected from the carboys used to store the raw waters and warmed to room temperature. The water was added to jars and floating covers were installed.

Then, lime alone or lime and soda ash were measured according to an experimental plan. For each lime dose, the required amounts of calcium hydroxide ($Ca(OH)_2$) were calculated as follows:

$$Ca(OH)_2\ (mg) = \text{Lime dose (mg/L CaO)} \times \frac{74\ \text{mg/mmol}\ Ca(OH)_2}{56\ \text{mg/mmol CaO}} \times 1L \qquad (3.1)$$

In several experiments, soda ash was also introduced to waters to achieve additional removals of calcium. As calcium precipitates with carbonate ion, carbonate ion becomes depleted, thus preventing further precipitation of calcium as calcium carbonate. Thus, carbonate ions are added as soda ash (Na_2CO_3) to allow continued calcium carbonate precipitation. For instance, Lake Austin water has the depletion point of carbonate at the lime dose of 125 mg/L; without adding carbonate at higher lime doses, the calcium concentration rises. The required amount of soda ash was determined based on the point that carbonate ions were limited. Therefore, the required amount of soda ash was calculated as follows:

$$Na_2CO_3(mg) = (\text{Lime dose}_{\text{desired}} - \text{Lime dose}_{\text{at depletion point of carbonate}}) \times \frac{1\ \text{mmol}\ Na_2CO_3}{1\ \text{mmol excess lime (CaO)}} \times \frac{106\ \text{mg/mmol}\ Na_2CO_3}{56\ \text{mg/mmol CaO}} \times 1L \qquad (3.2)$$

The measured mass of each chemical was mixed to make a slurry with 5 mL of distilled deionized water in a beaker. Using a 30-mL syringe, the slurry and an additional 5 mL of distilled deionized water that had been used to rinse the beaker was drawn and injected into softening jars. Separate syringes were used for soda ash, which was injected as a solution since soda ash is relatively easy to dissolve.

Rapid mixing was initiated at 150 rpm and the lime slurries were added to six jars simultaneously. Rapid mixing continued after lime addition for two minutes. Soda ash injection was conducted after one minute of lime addition during the rapid mixing. The rapid mixing lasted a total of three minutes in the case of soda ash addition.

In general, jar tests were performed at the standard condition: 2 or 3 minutes of rapid mixing depending on addition of soda ash, 30 minutes of slow mixing, and 30 minutes of settling. Supernatant was drained from the jars using the port that is 2.2 cm from the bottom.

Turbidity was measured directly, but all other analyses were preceded by filtration through 0.45 μm filter. Details of the softening jar test procedures are well documented in Ralls (1999).

3.3.2 Softening Batch Test

Softening as a pretreatment before ultrafiltration was conducted in a batch mode; a large amount of water (14 L) was softened in a 20 L cylindrical plastic reservoir at the specified softening condition. Since ultrafiltration was operated in a batch mode with a total recycle system (discussed below), the large amount of water was used to ensure that changes of water characteristics in the reservoir during membrane operation would be quite small.

The softening batch test was performed in the same way as the softening jar tests. It consisted of 2 minutes of rapid mixing (3 minutes of rapid mixing in the case of soda ash addition), 30 minutes of slow mixing, followed by a settling period. After 30 minutes of settling, a sample was taken for analysis, but a more extensive period of settling (up to several hours) was allowed before supernatant was siphoned off with a Tygon tube. The supernatant from the softening was used as the feed water for the subsequent ultrafiltration operation.

Filtration with a glass fiber filter was applied to simulate sedimentation in real treatment plants. Since turbidity is usually maintained around 3 to 5 NTU after sedimentation in real plants, the turbidity of settled water from the batch test was measured to obtain the same range of turbidity. If the water had a higher turbidity than 5 NTU, the water was filtered through a 1.5 μm glass fiber filter. Before use, the filter was prepared by rinsing with 500 mL of distilled deionized water. If the water had a lower turbidity than 3 NTU because of an extended settling during preparation, some sludge collected at the bottom of reservoir was added to obtain the desired range of turbidity. Immediately after the turbidity was adjusted, ultrafiltration was initiated using the water as a feed water.

3.4 ULTRAFILTRATION TEST

The ultrafiltration system used in this research was a bench-scale membrane apparatus with plate and frame module (Minitan-S, Millipore Company, MA) and was operated in a crossflow mode. A schematic diagram of the ultrafiltration system is shown in Figure 3.3.

The system had 30 cm^2 of filter area and had three ports: feed, filtrate, and concentrate ports. Pressure was measured by reading pressure gauges at the three ports. Flow rate was measured for filtrate and concentrate using a stopwatch and a graduated cylinder. Then, transmembrane pressure (TMP) was calculated from pressure measurements as below:

$$TMP = \frac{P_{feed} + P_{concentrate}}{2} - P_{filtrate} \qquad (3.3)$$

Since the pressure in the filtrate was atmospheric pressure, an average of the pressures in the feed and concentrate represented the transmembrane pressure.

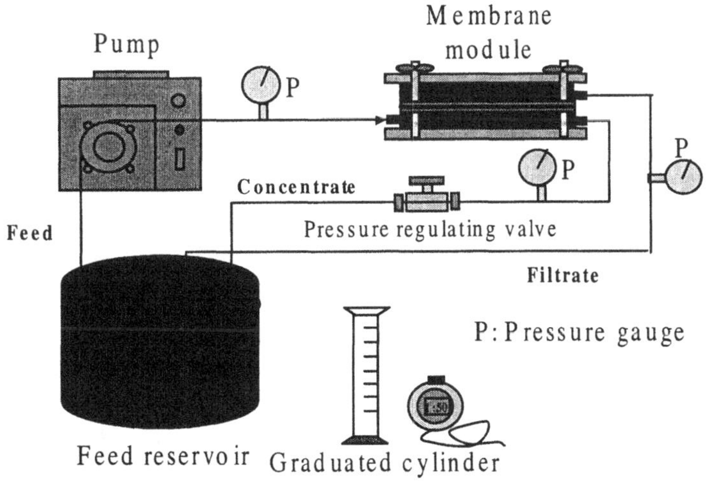

Figure 3.3 A schematic diagram of the ultrafiltration system: a total recycle system

3.4.1 Cleaning and Installation of the Ultrafiltration System

At the beginning of each test, a new membrane sheet was thoroughly cleaned to prevent DOC leach from preservatives used by the manufacturer. Cleaning procedures started with an overnight soaking of each sheet in a 0.5 N NaOH solution, which was recommended by the manufacturer. After rinsing with distilled deionized water, the membrane sheet was installed in the ultrafiltration system. Then, the membrane sheet was cleaned again two times with 2.5 L of distilled deionized water under conditions of approximately 70 kPa of TMP and 15 cm/s of clossflow velocity for 5 minutes. Lastly, at least 500 mL of filtrate were collected using a TMP of 105 kPa and a clossflow velocity of 2 cm/s, conditions at which the flow rate of filtrate was almost the same as the flow rate of concentrate. With these rigorous cleaning procedures, no dissolved organic carbon was detected in filtrate during the clean water specific flux measurement (described in the next section), and the membrane sheet, which was packed in a dry state, was totally wetted and stabilized with the distilled deionized water.

3.4.2 Clean Water Specific Flux

Clean water specific flux was measured with a cleaned and stabilized membrane sheet at a relatively constant pressure (i.e., approximately 90 kPa) for two hours. Only measurements from the last hour were used to calculate an average value of the clean water specific flux. During flux measurements, pressures at the feed and concentrate ports, filtrate flow rate, and temperature in the reservoir were observed and recorded every 10 minutes. The pressure gauges were written with the English unit, psi, and were converted to the S.I. unit, kPa, for this report. Usually, 3 or 4 mL of filtrate were collected in a graduated cylinder while the duration of that collection was monitored with a stopwatch. Since filtrate flow rates are the most important factors to indicate fouling, the flow rate measurement was repeated three times during the particular flux measurement, and the average calculated. The filtrate flow rate was converted to the specific flux by dividing by the membrane area (i.e., 30 cm^2) and applied transmembrane pressure, as follows:

$$J = \frac{V(mL)}{t(sec)} \cdot \left(\frac{L}{1000mL} \cdot \frac{3600sec}{hr} \right) \Big/ A\,(m^2) \cdot P\,(psi) \left(\frac{100\,kPa}{14.5\,psi} \right) \tag{3.4}$$

in which J is the specific filtrate flux (L/m^2-hr-kPa), V is the average of measured filtrate volumes, t is the collecting time, A is the membrane area (*i.e.*, 0.003 m^2), and P is the transmembrane pressure.

The clean water specific flux was reported after the flux was corrected for temperature (i.e., normalized to 20 °C) because the membrane operation was conducted at room temperature and there was a minor variation of a few degrees at various times of the year. The equation for temperature correction by Braghetta et al., (1997) was used as follows:

$$J_{at\ 20°C} = J \cdot \exp\left(-0.024\,(T°C_{measured} - 20°C)\right) \tag{3.5}$$

in which T is the water temperature in the feed reservoir. The equation accounts for the change in the viscosity of water with temperature. The clean water specific flux represents the flux through the membrane with a resistance only from the membrane itself; thus, it is the flux before any fouling occurs.

3.4.3 Ultrafiltration

After measuring the clean water flux, the ultrafilter was run with a feed water, which was either raw water without softening or a softened water. Each experiment was performed in a batch mode, with softening of the water followed by ultrafiltration. The duration of each membrane experiment varied depending on the pattern of flux decline. However, it lasted until the normalized cumulative production (cm^3/cm^2), which is the volume of water passed through a unit membrane area in a standard membrane sheet (discussed subsequently), was at least 60 cm^3/cm^2. Ultrafiltration was operated as a total recycle system, i.e., waters from both concentrate and filtrate ports were returned to the feed reservoir. The water volume in the feed reservoir (14 L) was significantly larger than that in the ultrafiltration system (approximately 30 mL). Therefore, the small mass of materials deposited or adsorbed on the ultrafilter would yield little change in the water quality of the feed reservoir. The interval for flux measurements was every 5 minutes at the beginning, and then increased to 30 minutes toward the end of the experiments. Flux measurements were performed the same way as the clean water flux measurements, i.e., pressure measured in the feed and concentrate ports, the flow rate of filtrate measured three times, and temperature in the reservoir recorded. The specific water flux for samples was also calculated as discussed in the Section 3.4.2. Then, the normalized specific flux (i.e., percent of the clean water specific flux) was used as a method to compare flux decline behavior with other membrane tests.

At the conclusion of each experiment, membranes were either washed with a series of cleaning methods or frozen with liquid nitrogen and then dried in a freeze dryer for further membrane surface analyses.

3.4.4 Cleaning Methods for Fouled Membranes

Cleaning methods for fouled membranes included a surface wash with distilled/deionized water, a caustic wash with a high pH solution (i.e., 0.5 N NaOH), and an acidic wash with a low pH solution (i.e., 0.1 N HNO_3). The caustic solution was intended to remove organic foulants, and the acidic solution was to remove inorganic precipitates. After the water in the ultrafiltration apparatus was withdrawn, 250 mL of cleaning solution served as the feed water and was recirculated. The cleaning solution was recycled at the highest speed of the pump (Variable Speed Tubing Pump, Millipore Company, MA.), i.e., 480 mL/min of flow rate, which yielded a crossflow velocity of approximately 16 cm/s. The duration of each cleaning was one hour for the surface wash, two hours for the caustic wash, and one hour for the acidic wash. After each cleaning, the cleaning solution was emptied from the system and the clean water specific flux was measured again using distilled/deionized water for 20 minutes. The clean water flux measurements after each cleaning method were to evaluate the efficiency of each cleaning process and thereby learn indirectly about the causes of fouling.

3.5 LIQUID SAMPLE ANALYSIS

After softening and solid-liquid separation, each liquid sample was characterized to determine the removal of inorganics and organic matter in softening. In addition to samples from softening, liquid samples from membrane processes such as feed, filtrate, and concentrate were also characterized.

Most samples were analyzed without filtration. The turbidity of feed waters after softening was adjusted to be approximately 3-5 NTU to simulate plants in practice. However, in cases in which high turbidity waters were applied to ultrafiltration, feed and concentrate waters were filtered through a 0.45 μm membrane filter (Type HA, Millipore Co.). To eliminate leaching of organic carbon from the filter into samples, the filter was rinsed with 500 mL of distilled/ deionized water prior to use. Preliminary tests confirmed that the DOC after this procedure was below the limits of detection.

The liquid sample analyses were performed as shown in Table 3.4. Ralls (1999) thoroughly described the analyses of pH, calcium, magnesium, DOC, and UV absorbance; most of these details are repeated in Appendix A. During the research, instruments for DOC and UV absorbance analyses were changed as described in Smith (2001). Quality assurance tests were performed on several samples to ensure that measurements between the analyzers were consistent.

Several analytical methods such as turbidity, hydrophobic DOC, and size-exclusion chromatography are explained in detail later.

3.5.1 Turbidity

Turbidity was measured with a Hach Ratio/XR turbidimeter. At the beginning of this research, the meter was calibrated with the primary standard solutions, *i.e.*, 1.8, 18, 180, and 1800 NTU of formazin solution (Hach Co.) and measured turbidity of the secondary standard solutions (Gelex Secondary Turbidity Standards: 0~2, 2~20, 20~200, and 200~2000 NTU). The recorded turbidity of the secondary standards at the primary calibration was used as an indicator when the next calibration with the primary standard solutions was necessary.

Table 3.4
Analytical Methods

Analysis	Constituent	Comments
Softening	pH	Orion Research Model 701A with Orion sureflow Ross Combination pH probe
	Alkalinity	Titration with 0.01N H_2SO_4 until the carbonic acid endpoint was reached
	Calcium	Perkin Elmer 2380 Flame Atomic Absorbance (FAA) Spectrophotometer
	Magnesium	Perkin Elmer 2380 FAA Spectrophotometer
	Turbidity	Hach Ratio/XR turbidimeter
NOM	UV absorbance	Perkin Elmer Lambda 38 5 cm cuvette or Agilent 8453E UV-visible Spectroscopy
	Organic Carbon	Dohrman DC-180 TOC analyzer and Tekmar- Dohrman Apollo 9000 TOC Combustion Analyzer with a STS 8000 Autosampler
	Hydrophobic DOC	Extraction with XAD-8 and DOC analysis with the Tekmar-Dohrman Apollo 9000 TOC Combustion Analyzer with a STS 8000 Autosampler
	SEC*	Waters SEC with Ultra Hydrogel™ Linear column and a differential refractometer R401 detector

* SEC: Size-exclusion chromatography

The filtrate from UF membranes has very low turbidity. In the early part of the experimentation, some samples of UF filtrate might have been slightly contaminated because samples were collected in a beaker and then poured into the vial for the turbidimeter. Later, filtrate samples flowed directly from the tube of the membrane system into the vial. After this change, all such samples had a measured turbidity of 0.05 NTU or less, which is essentially the lower limit of detection of this instrument. Because of the change in procedure and the limits of the instrument (at least as used in this research), differences in the filtrate turbidities are insignificant.

3.5.2 Hydrophobic DOC

Organic carbon in several waters was fractionated to determine their hydrophilicity using an XAD-8 resin adsorption procedure (Thurman and Malcolm 1981, Singer et al. 1995).

The XAD-8 resin (Supelite DAX) was purchased from Supelco (Bellefonte, PA) and thoroughly cleaned according to the procedures of Thurman and Malcolm (1981). The cleaning included a 5-day rinse with 0.1 N NaOH and Soxhlet extraction for 24 hrs with methanol, diethyl ether, acetonitrile, and methanol. Then, 10 mL of the pre-cleaned XAD-8 resin was packed into a 1-cm diameter glass column and was rinsed sequentially with distilled/deionized water, NaOH, and HCl solutions. The detailed procedures were thoroughly described in Singer et al., (1995).

Samples were filtered through 0.45 μm membrane filters, which were cleaned with 500 mL of distilled/deionized water to ensure no DOC leached from the membrane itself. After acidifying to pH 2 with concentrated HCl, samples were run through the column at a flow rate of 15 bed volumes per hour, i.e., 2.5 mL/min for the column in the experiments. Organic carbon was measured in both influent and effluent samples. The fraction of the organic carbon adsorbed by the XAD-8 resin was calculated and reported as the hydrophobic DOC for the samples.

For Lake Austin water, hydrophobic DOC adsorption was performed on raw water, softened waters at three levels of softening, and filtrate from ultrafiltration. For Missouri River water, raw water and softened waters at 90 and 165 mg/L lime doses were taken for measurements.

3.5.3 Molecular Weight Distribution – Size Exclusion Chromatography

Size exclusion chromatography (SEC) has been used to determine molecular size (or weight) and molecular weight distributions in addition to separation and quantitation of samples (Buffle 1990). In SEC, a solution is injected into a solvent stream, which then flows through a column or series of columns to be separated by size or molecular weight. Columns are packed with a porous gel, which contains pores with comparable sizes to molecules to be separated. Organics with high molecular weights pass through the columns faster than smaller organics. Therefore, the molecular weight distribution can be analyzed by measuring the time of departure from the column. The organics in the eluted solution from the columns are measured by a concentration detector such as a differential refractometer, a UV absorbance detector, or an infrared flow-through spectrometer (Handley 1993).

Raw and softened waters from Lake Austin and dextran were analyzed for molecular weight distribution. Water samples were concentrated by a vacuum rotary evaporator at 30°C to obtain a DOC between 300 and 400 mg/L C. One hundred microliters of the concentrates were injected into two columns (Ultrahydrogel Linear, Waters) in series and eluted with a 0.1 N

$NaNO_3$ solution at a rate of 1 mL/min. Detection of the eluent was measured using a differential refractometer (R401, Waters), which detects differences in the refractive index of the eluent between a sample and a reference stream.

3.6 MEMBRANE SURFACE ANALYSIS – SEM

Membrane surface analysis with scanning electron microscopy (SEM) was performed to investigate fouling on the surfaces of membranes. Images from SEM enable comparison of surfaces of clean and fouled membranes and give qualitative proof of deposition and adsorption of foulants. Two SEM instruments were available for this research, a Hitachi-S 4500 field emission scanning electron microscope (FESEM) and a JEOL T330A SEM. Through several preliminary analyses of membrane samples, it was revealed that the JEOL instrument offered an easier and more precise way to acquire images for samples; thus the JEOL T330A (Jeol USA. Inc., Peabody, MA) was used for this research.

3.6.1 Sample Preparation – a Freeze Drying

To prevent interference by evaporating water under high vacuum in the SEM chamber, all membranes for SEM analysis must be completely dried. Several drying methods have been used: the air drying method, the critical point drying method, and the freeze drying method. The latter method was selected because it preserves sample specimens, even organic matter, relatively well. Immediately after ultrafiltration, a portion of the membrane sheet was cut into three parts and soaked in liquid nitrogen to freeze the present status of the ultrafilter. The frozen membrane sheets were then put in a Pyrex tube and installed in the freeze dryer, which maintains the temperature below –60 °C and the vacuum under 80 millitorr. The freeze dryer was operated overnight to ensure complete drying of samples.

3.6.2 Sample Preparation – a Gold Coating

All samples for SEM must have a conductive coating applied to their surfaces. For membrane samples in this research, a gold coating and sometimes a gold/platinum alloy were applied.

After samples were mounted on 1-cm diameter Al-cylinders with glue, they were generally coated for 40 seconds at a current of 40 mA. A shorter time for coating, *i.e.*, 30 seconds, was applied to samples for high resolution.

3.6.3 SEM Operation

Images of membrane sheets were usually taken under an acceleration voltage of 15 kV, which allowed magnification in the range of 1k to 100k. Lower acceleration voltages at 1, 2, and 5 kV were tried to obtain a higher resolution, such as 300k or 500k magnification, but these procedures resulted in little success. In general, 2k or 10k of magnification was used to investigate overall images of fouling layers on membrane surfaces. The high resolution, *i.e.*, higher than 200k magnification, was attempted (with mild success) to obtain detailed images of pores.

3.7 MEMBRANE SURFACE ANALYSIS – XPS

The composition of the outermost few nanometers of the membrane surface was measured by X-ray photoelectron spectroscopy (XPS). XPS has been used to identify and quantify the elemental and functional groups within the surface.

Sample preparation was the same as for SEM analysis, i.e., freeze drying overnight. Then, the XPS data were obtained using a Physical Electronics PHI5700 ESCA system equipped with an Al monochromatic source (Al Kα radiation at 1486.6 eV). The base pressure in the XPS ultra-high vacuum (UHV) chamber was 1×10^{-10} torr. Wide range (survey) scans were obtained with a step size of 1 eV and pass energy of 93.9 eV; high resolution scans were taken with a step size of 0.1 eV and pass energy of 11.75 eV. The Ag3d5/2 XPS peak at 368.3 eV from a sputtered-clean Ag foil was used to calibrate the system.

In most cases, the measured atomic composition by the high resolution scan was for carbon, oxygen, sulfur, calcium, and magnesium (C, O, S, Ca, Mg) of the clean and fouled membranes. When Lake Austin water was used as a feed water, the atomic composition of silicate (Si) was additionally scanned to examine its role in membrane fouling.

3.8 SUMMARY

The experimental procedures were designed to investigate the feasibility of softening as a pretreatment for ultrafiltration. Softening was performed at two or three lime doses, which represented different levels of softening pretreatment to UF in terms of organic matter (i.e., NOM fouling) and inorganics (i.e., inorganic fouling by further precipitation). Effects of softening pretreatment on UF were investigated with flux decline behavior during ultrafiltration and recoveries of the clean water specific flux by three different cleaning methods afterwards. Each cleaning method was related to a particular fouling mechanism: a surface wash to fouling by particle deposition, a caustic wash to organic fouling, and an acidic wash to inorganic fouling by precipitation. With measurements of clean water specific flux after each cleaning method, the relative extent of fouling mechanisms could be evaluated. Also, sophisticated instruments such as SEM and XPS were applied to understand fouling phenomena directly from the surface of membranes.

CHAPTER 4
RAW WATER CHARACTERISTICS AND SOFTENING CONDITIONS

The presentation of experimental results is divided into two chapters. The first chapter includes characteristics of the two natural waters chosen for this study and the determination of the specific softening conditions to use as a pretreatment for ultrafiltration. The softening conditions of the two natural waters were selected based on lime doses, i.e., the degree of softening, which determines the organic matter and particle concentration of the feed water to ultrafiltration. In addition, several scenarios of softening were investigated to discern a process with little fouling. The softening processes included lime softening alone, lime softening with pH adjustment, lime-soda ash softening, and lime-soda ash softening with pH adjustment. In addition, this chapter introduces a method to normalize flux and set a standard operational period for the laboratory experiments. The next chapter presents results of fouling phenomena when softening is used as a pretreatment for ultrafiltration.

4.1 RAW WATER CHARACTERISTICS

Two natural water sources, Lake Austin water (Austin, TX) and Missouri River water (St. Louis, MO) were selected based on Thompson et al. (1997) and work in our laboratory. The water quality characteristics of these waters are reasonably similar except for turbidity.

To characterize the source waters, the key water quality parameters of the two natural water sources are listed in Table 4.1. The turbidity results are substantially different but the other water quality parameters such as TOC and Mg^{2+}: Ca^{2+} ratio are rather similar to each other. Both hardness ions, Ca^{2+} and Mg^{2+}, are slightly lower in the Missouri River water than in Lake Austin water. Due to the dramatic difference in particle concentrations, Missouri River water is useful to investigate effects of particle concentration on fouling in ultrafiltration. As noted in Chapter 2, particulate fouling is a major potential problem in ultrafiltration, and is therefore a primary interest in this research.

The Mg^{2+}: Ca^{2+} ratio can affect the amount of formation of magnesium hydroxide, which is thought to have an important role in the removal of NOM in enhanced softening. Calcium carbonate has a limited capacity to adsorb NOM due to a relatively small surface area caused by its compact structure and the repulsive interaction of its negative charge with the similarly charged NOM. On the contrary, $Mg(OH)_2$ precipitate has a very fluffy shape (providing a high surface area) and a positive surface charge, both of which help to adsorb NOM onto the solids. Also, SUVA values are an indicator of the hydrophobic portion of NOM; generally SUVA values above 3.0 L/m-mg indicate a water with a substantial portion of hydrophobic NOM. Hydrophobic organics have been cited as the most easily removed organic matter in enhanced softening (or enhanced coagulation) (Randtke *et al.* 1994), but are also the major cause of fouling material in membrane processes (Nilson and DiGiano 1996). Both of these waters have relatively low SUVA values, but they are still sufficient to allow adequate testing of softening pretreatment to reduce organic fouling of membranes.

4.2 EXTENT OF SOFTENING

The degree to which the water can and should be softened is different for waters from different sources. In some waters, the degree to which the water should be softened is quite obvious, if softening alone is the measurement of performance. For instance, in a certain pH (or lime addition) range for Lake Austin water, very little in terms of hardness (calcium) removal was accomplished despite further addition of lime over a substantial range of doses. For other waters, there is an essentially continuous fall in the hardness with increasing lime addition (until essentially all hardness is removed). Furthermore, the amount of lime added (for a given water) dictates the amount of TOC removed in enhanced softening.

Table 4.1
Water quality of two water sources

Water quality parameter	Water source	
	Lake Austin Austin, TX	Missouri River St. Louis, MO
pH	8.3	8.1
DOC (mg/L C)	4.2	4.1
Ca^{2+} (mg/L $CaCO_3$)	154 (61.7*)	133 (53.3*)
Mg^{2+} (mg/L $CaCO_3$)	80 (19.4*)	63 (15.3*)
Mg^{2+}: Ca^{2+}	0.52	0.47
Alkalinity (mg/L $CaCO_3$)	179	129
UV_{254} (cm^{-1})	0.097	0.087
SUVA (L/m-mg TOC)	2.3	2.1
Turbidity (NTU)	1.8	314
Hydrophobic DOC (%)	44	32

(*): reported in units of mg/L

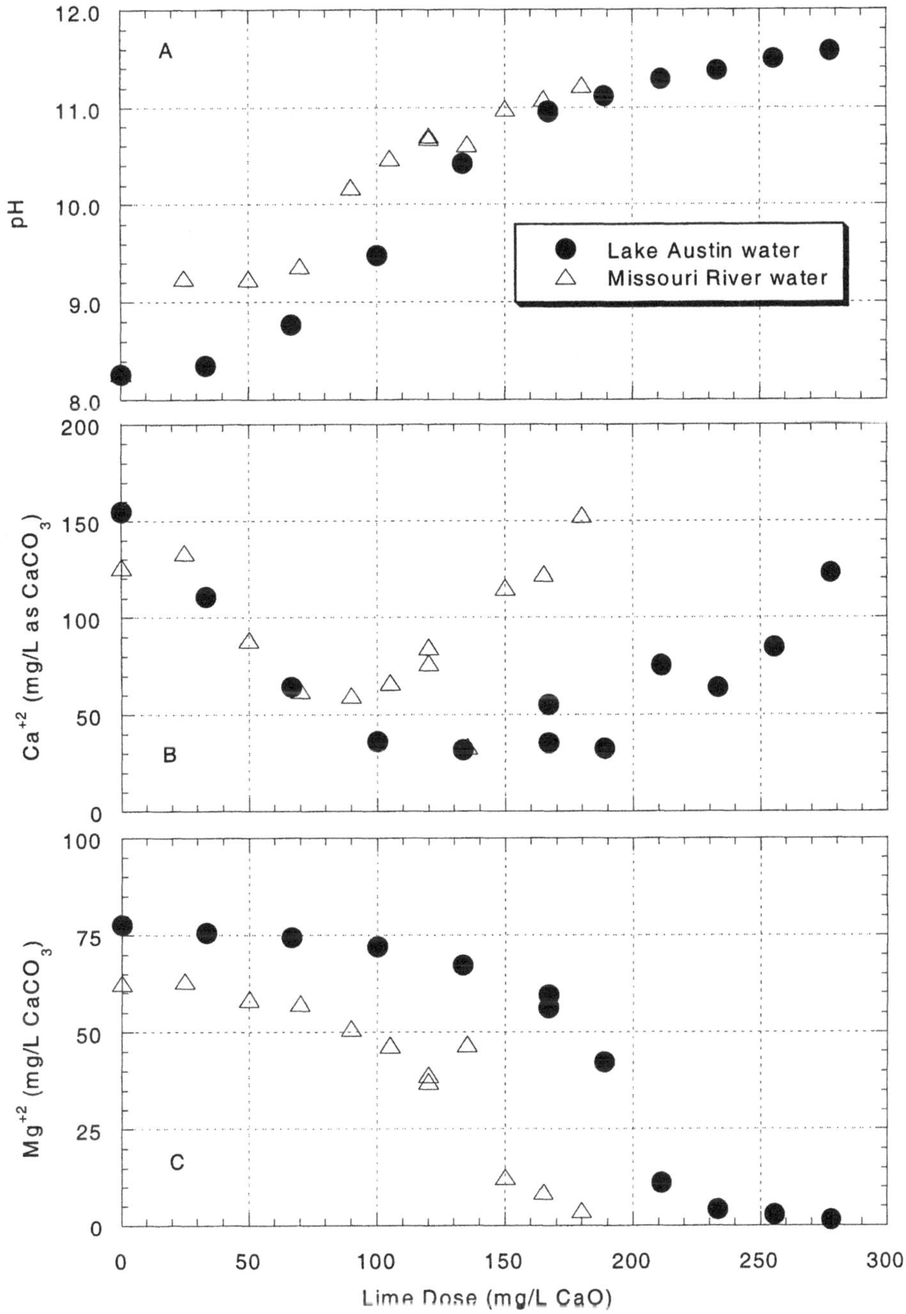

Figure 4.1 Softening results for Lake Austin and Missouri River

The magnesium concentrations (Figure 4.1C) in both waters decreased slightly for the first several lime doses (presumably by co-precipitation or adsorption of Mg^{2+} ions into the $CaCO_3$ precipitate), and then decreased dramatically by $Mg(OH)_2$ precipitation. However, the $Mg(OH)_2$ precipitated at a lower lime dose in the Missouri River water due to the lower initial alkalinity concentration (i.e., more rapid rise in pH).

The removal of natural organic matter (NOM) is also a function of the amount of lime added, as indicated in Figure 4.2. Both direct measures of NOM (i.e., the dissolved organic carbon and the UV absorbance at 254 nm) decreased continuously with increasing lime dose in both waters. Interestingly, the specific UV absorbance (SUVA), an indicator of the hydrophobic content of the remaining DOC, also decreased, meaning that the hydrophobic portion of the NOM was selectively removed by the enhanced softening, as one would expect.

4.2.2 Selection of Lime Doses for Each Water Source

In this research, we selected three levels of softening conditions for each water: a "standard softening" condition chosen on the basis of softening parameters alone (i.e., minimum lime dose to achieve excellent calcium removal), a second "enhanced softening" condition that would represent the maximum lime dose without the voluminous precipitate of $Mg(OH)_2$, and a third "Mg softening" condition at a higher lime dose where the precipitation of $Mg(OH)_2$ would occur to a significant extent. A softening plant concerned about softening alone would likely operate at the first condition. A softening plant using enhanced softening would likely operate at the second condition (or would consider it as the upper bound of the conditions that might be used). A plant that treats a water with a substantial Mg^{2+} concentration but that was designed for Ca^{2+} removal only would likely have serious operational problems if they exceeded this dose because the subsequent solid-liquid separation units (sedimentation and filtration) would not have been designed to handle the voluminous, low-density $Mg(OH)_2$ sludge. Some plants might be forced to go to $Mg(OH)_2$ precipitation for maximum NOM removal, and this would be represented by the third condition. The pH or lime dose to accomplish these conditions would vary with water sources.

Therefore, in a plant with ultrafiltration following precipitative softening, the degree of pretreatment (i.e., the extent of softening) determines the organic matter and particle concentration of feed water to ultrafiltration, and thus might influence the degree of fouling and the success or efficiency of membrane treatment.

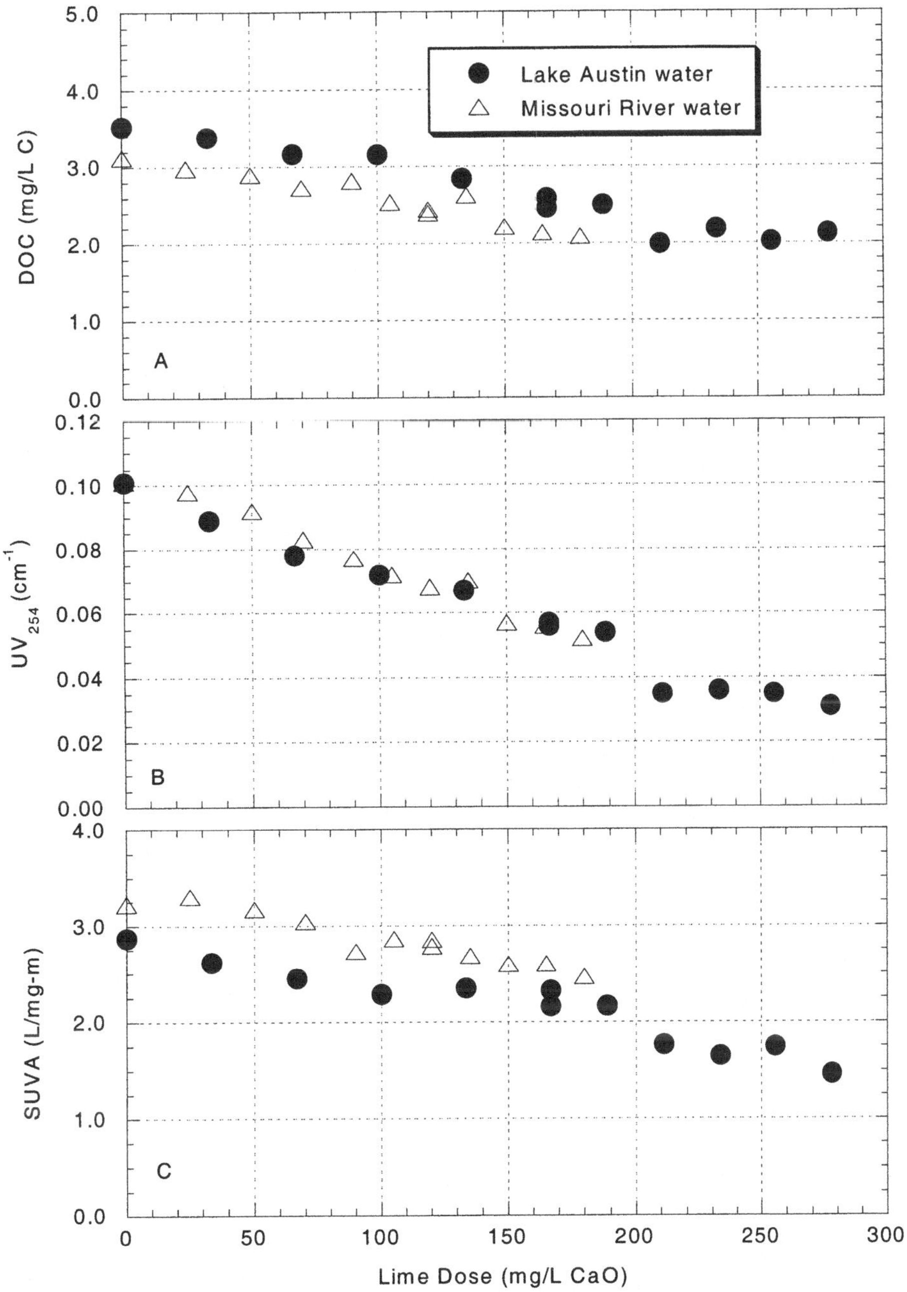

Figure 4.2 Enhanced softening results for Lake Austin and Missouri River

Based on the results for Lake Austin water presented in Figure 4.1, the three lime doses for Austin water are selected to be 125 mg/L as CaO for standard softening (the dose where the minimum Ca^{2+} is first reached), 170 mg/L for enhanced softening (where $Mg(OH)_2$ is just beginning to precipitate), and 230 mg/L for Mg softening (where $Mg(OH)_2$ precipitation is almost complete). The NOM (i.e., DOC) removals achieved at these three lime doses were 11%, 29%, and 37%, respectively; we expect, therefore, that the amount of fouling that will occur during operation of ultrafiltration would be different at the different lime doses. The key water properties of Lake Austin water for each condition are summarized in Table 4.2.

The standard condition represents the optimum operation condition when only softening is considered. Therefore, the calcium concentration is the least, 31 mg/L as $CaCO_3$ at 125 mg/L of lime dose, but the magnesium concentration is reduced only slightly.

At the enhanced softening condition, 170 mg/L of lime dose, the organic matter (i.e., DOC) is removed to a much greater extent than that at the standard condition, even though the alkalinity is changed only slightly due to the compensating effects of increasing calcium and decreasing magnesium. As shown in Figure 4.1C, the magnesium hydroxide starts to precipitate at approximately the 170 mg/L of lime dose. Therefore, we can get the highest removal of NOM without creating the sludge problem from the voluminous $Mg(OH)_2$ precipitate at the enhanced softening condition.

Further organic matter removal, i.e., 37% DOC removal, occurred at the higher lime dose, the Mg softening condition, i.e., 230 mg/L of lime. With this higher lime dose, the NOM fouling should be less due to the lower concentration of DOC, but the particle fouling would be greater due to the higher concentration of particles in the unlikely event that ultrafiltration was used without prior sedimentation or deep bed filtration.

The three softening conditions for the Missouri River water also were studied, but both the standard and the enhanced softening conditions yielded a dose of 90 mg/L of CaO because $Mg(OH)_2$ started to precipitate around the same dose that calcium had the lowest concentration. Therefore, 90 mg/L as CaO was selected for both the standard and the enhanced softening conditions, and 165 mg/L of lime was chosen for the Mg softening condition for the Missouri River water. The NOM (i.e., DOC) removals achieved at these two lime doses were 10% and 32%, which showed significant NOM removal at the Mg softening condition. The key water properties of the Missouri River water for each condition are summarized in Table 4.3.

4.3 INVESTIGATION OF SOFTENING SCENARIOS

Softening pretreatment before ultrafiltration could be practiced in several ways in real plants, *i.e.*, lime softening alone, lime softening with pH adjustment (re-carbonation), lime/soda ash softening, or lime/soda ash softening with pH adjustment. The question for this research is which softening process should be used to reduce fouling in ultrafiltration.

Depending on performance in the softening process, the constituents remaining in softened water such as solids, hardness ions, and NOM concentrations can affect fouling in ultrafiltration as particulate, inorganic, or organic fouling, respectively. To separate the inorganic and organic fouling, a synthetic water with only inorganic constituents was used to investigate inorganic fouling exclusively. This water mimics the inorganic constituents of Lake Austin water. With the synthetic inorganic water, the NOM effects on both softening and membrane performance are avoided.

Various scenarios for softening pretreatment were tested using the water with only inorganic constituents, and then with Lake Austin water. The synthetic inorganic water was used to find a specific softening process with the least inorganic fouling on the membrane surface. In addition, effects of NOM and particles on ultrafiltration with different scenarios of softening pretreatment were investigated with Lake Austin water. Then, the specific softening process with the least fouling was used in the subsequent experiments.

Table 4.2
Softening and enhanced softening results of Lake Austin water for each chosen lime dose

	Standard softening condition	Enhanced softening condition	Mg softening condition
Lime (mg/L as CaO)	125	170	230
pH	10.3	11.0	11.4
Ca^{2+}(mg/L as $CaCO_3$)	31	44	64
Ca^{2+}(mg/L)	12.4	17.6)	25.7
Mg^{2+}(mg/L as $CaCO_3$)	65	58	4.2
Mg^{2+}(mg/L)	15.8	14.1	1.0
DOC (mg/L as C)	3.1	2.5	2.2
SUVA (L/mg-m)	2.35	2.24	1.65

Table 4.3
Softening and enhanced softening results of the Missouri River water for each chosen lime dose

	Standard and Enhanced softening condition	Mg softening condition
Lime (mg/L as CaO)	90	165
pH	10.2	11.1
Ca^{2+}(mg/L as $CaCO_3$)	59.9	123
Mg^{2+}(mg/L as $CaCO_3$)	51.0	9.0
DOC (mg/L as C)	2.8	2.1
SUVA (L/mg-m)	2.74	2.62

4.3.1 Normalization of Flux and Operational Period

Prior to presentation of results from ultrafiltration, this section describes a method to normalize flux and set a standard operational period for the laboratory experiments. The normalization appropriately accounts for variations in pore area and transmembrane pressure (TMP) between experiments in the laboratory scale system.

Hydraulic Water Flux

Flux decline has been used to indicate the extent of fouling, such that a rapid flux decline means severe fouling. Water flux can be expressed in several ways, including the specific flux and percent of the clean water specific flux.

The specific flux is the flux divided by the applied transmembrane pressure (TMP). This normalization is valuable because water production (i.e., flux) increases with TMP. If a system is operated at constant pressure, the decline of the specific flux is identical to that of the water flux. However, it is convenient to compare results from experiments with different TMP by normalizing flux with the applied TMP.

Another method to normalize water flux is to express it as the percent of clean water specific flux, which is especially useful in laboratory scale experiments. The clean water specific flux is the specific flux of the distilled/deionized Milli-Q (Millipore Company, Bedford, MA) water of a clean membrane. Even though membranes are made from the same material by the same manufacturer and have the same nominal molecular weight cut-off (MWCO), each membrane sheet has a different initial flux due to different pore densities in the various sheets. Therefore, the decline of the specific flux in a certain experimental condition is monitored relative to the initial specific flux from the clean Milli-Q water. With the percent of clean water specific flux, we can compare the flux decline behaviors in different experiments based on the initial specific flux from the clean water.

$$\%\text{ of clean water specific flux} = \frac{\text{Specific flux}_{\text{at certain condition}}}{\text{Clean water specific flux}} \times 100 \qquad (4.1)$$

Operational Period

For the abscissa of the flux decline behavior of a membrane system, the elapsed time such as minutes or hours has often been used for the operational period. However, within the same time period, a UF sheet with higher pore density has a higher flow through the membrane than one with lower pore density. The amount of flow through the membrane is directly related to fouling. Therefore, a cumulative water production (the permeate throughput/ membrane area, i.e., cm^3/cm^2 or $m^{\square}/m^2$), instead of the elapsed time, appears more reasonable to represent the operational period. The cumulative water production shows the flux decline based on the volume of water treated by a unit area of membrane. The cumulative water production can provide a general way to compare results from other experimental conditions and even from real plants. With the elapsed time, it is hard to obtain the water production by the specific membrane without detailed information on water flux and available membrane areas.

In addition, the water production from each membrane sheet with the same nominal membrane area could be variable due to the differences in pore areas on membrane sheets.

Therefore, a standard condition of pore area in the membrane is set to be 5% porosity. The water flux of this standard UF sheet is calculated assuming Poiseuille flow in the pore as shown below (Wiesner and Aptel 1996).

$$J = \frac{f \cdot r^2 \cdot \Delta P}{8 \cdot \mu \cdot \theta \cdot \delta_m} \tag{4.2}$$

J: flux (L/m^2-hr); f: the fraction of open pore area on the membrane surface (=0.05); r: radius of the pore (=0.01 μm); ΔP: applied TMP (=90 kPa); μ: viscosity of water at 20 °C (=1 cp); θ: tortuosity (=1); and δ_m: effective thickness (=1.4 μm).

Since it is hard to measure pore density of each UF sheet, the specific flux of the clean water was used in this research as an indicator of its pore area. Therefore, the ratio of pore area of the standard sheet to each UF sheet is assumed to be the same as the ratio of the specific flux of clean water from the standard sheet to each UF sheet. The water production normalized by the ratio of pore area of the standard sheet to each UF sheet leads to the normalized cumulative production (NCP).

$$\text{Normalized cumulative production } (cm^3/cm^2) = \frac{\text{permeate throughput}}{\text{membrane area}} \cdot \frac{\text{clean water specific flux}_{\text{for standard UF sheet}}}{\text{clean water specific flux}_{\text{for each UF sheet}}} \tag{4.3}$$

To illustrate the value of the normalization of both the ordinate and abscissa, consider the data shown in Figure 4.3. Two data sets for flux decline showed a big difference when represented with the specific flux as the ordinate and minutes as the abscissa although they were obtained under the same conditions (*i.e.*, lime dose of 125 mg/L as CaO, polysulfone membrane, and a pore size of 10 kDa MWCO). However, the same two data sets overlap closely after the normalizations described above were performed on both the ordinate and the abscissa. The normalizations of the specific flux and operational period are required to overcome the variations in pore area and TMP in the laboratory scale systems. Small variations in large-scale systems have little effects on water flux decline because the enormous membrane area eliminates effects of the variations of pore density in small areas of membrane. Hence, this normalization of the abscissa values to account for differences in membrane sheets is probably only important for laboratory scale work. Because all experiments in this research were performed in a laboratory scale system, the normalizations were used throughout this research.

The normalization of the abscissa and ordinate accounts for differences between membrane sheets but do not otherwise hide differences between experiments. Duplicate experiments for several conditions were performed throughout this research, and the flux measurements (and water quality results) for these duplicates are reported in Appendix B.

4.3.2 Softening Scenarios with Synthetic Inorganic Water

Several variations of the softening process were investigated with the synthetic inorganic water to discern a process with little inorganic fouling. The softening processes included lime softening alone, lime-soda ash softening, and lime-soda ash softening with pH adjustment.

Carbon dioxide was used to reduce the pH after softening into the range of 8.5 to 9.5. The soda ash was applied to increase the calcium ion removal at the high lime doses of 170 mg/L and 230 mg/L as CaO; without soda ash addition, the calcium concentration after softening is higher than at the lower dose of 125 mg/L because the carbonate concentration is so low that it limits further $CaCO_3$ precipitation.

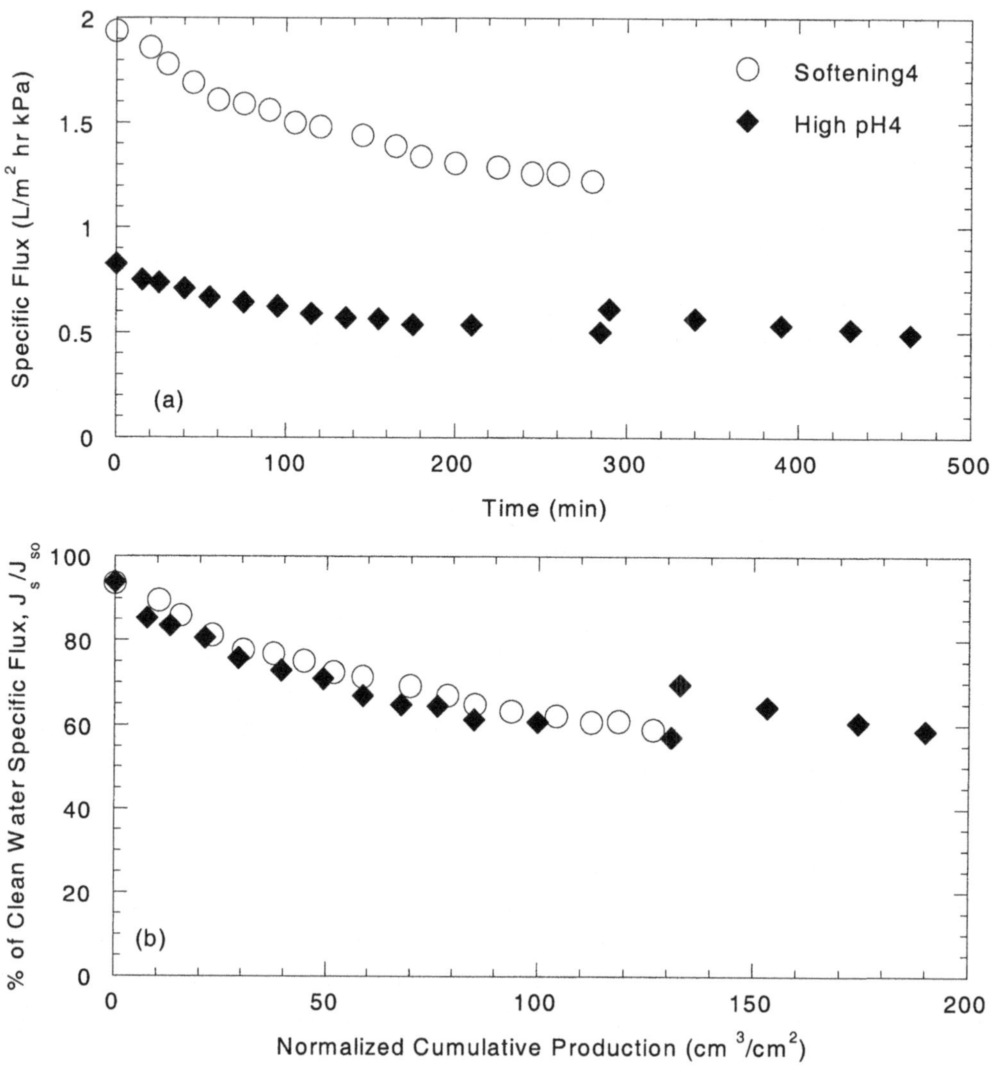

Figure 4.3 Example of the normalization. (a) before and (b) after normalization (Both experiments with the following conditions: lime dose of 125 mg/L CaO, and polysulfone membranes with MWCO of 10,000)

The experiments to investigate the value of soda ash were conducted at a lime dose of 230 mg/L. The dose of soda ash was based on the softening results from experiments with lime addition only. The soda ash dose (in equivalents per liter, or other comparable units) was the difference between the chosen lime dose for each experiment and the lime dose that yielded the least calcium concentration without any soda ash addition. The lime dose that showed the least calcium concentration was 125 mg/L.

The operational conditions are presented in Table 4.4. TMP was maintained around 90 kPa and the crossflow velocity was approximately 10 cm/s. The clean water specific flux also fluctuated in these experiments, apparently due to substantial differences in the membrane sheets. In this table and throughout the body of this report, the results reported are from experiments using the polysulfone membrane; results from the few experiments with the cellulose acetate membrane are shown in Appendix C.

Table 4.5 shows the water quality at various stages of the experiments. The high values in Ca^{2+} concentration in the softened water for all three experiments were due to both soluble calcium and $CaCO_3$ particles that were not settled, which was also reflected in the high turbidity values for all experiments. The experiment with lime softening only showed that calcium concentration was quite high in the feed and filtrate. Since carbonate ion was already depleted at the lime dose of 125 mg/L, the added lime above that level increased the calcium concentration. With soda ash addition in the other two experiments, the calcium was removed to lower than 5 mg/L of Ca^{2+} in the UF feed water. Magnesium was almost completely removed by $Mg(OH)_2$ precipitation at the lime dose of 230 mg/L.

Table 4.4
Operational conditions for each test with synthetic inorganic water at 230 mg/L of lime doses using the different softening processes

Lime (mg/L CaO)	Soda ash (mg/L CaO)	pH adjustment	TMP (kPa)	Crossflow velocity (cm/s)	Clean water specific flux, J_{so} (L/m^2-hr-kPa)
230	-	No	94.5-96.6	10.4	1.68
230	105	No	88.3-92.4	10.2	1.87
230	105	Yes	88.3-94.5	10.3	2.20

Table 4.5
Water quality of experiments with synthetic inorganic water at 230 mg/L of lime doses using the different softening processes

Lime dose (mg/L)	Soda ash dose (mg/L)	pH adj.	Sample	pH	Turbidity (NTU)	Ca^{2+} (mg/L)	Mg^{2+} (mg/L)
230	0	No	Raw	8.54	-	57.6	19.1
			Softened	11.81	171.1	77.8	9.3
			Feed	11.67	5.05	43.5	3.2
			Filtrate	11.64	0.055	35.6	0.3
230	105	No	Raw	-	-	-	-
			Softened	12.00	59.7	22.8	2.4
			Feed	11.91	2.76	1.8	0.5
			Filtrate	11.95	0.027	1.5	0.3
230	105	Yes	Raw	8.42	-	46.5	17.4
			Softened	12.01	88.0	35.2	0.7
			Feed	8.89	2.52	4.6	0.8
			Filtrate	8.87	0.011	4.5	0.8

The membrane results for these experiments at 230 mg/L lime dose, shown in Figure 4.4, indicated that the flux decline was insignificant regardless of which softening process was used during 120 cm^3/cm^2 of normalized cumulative production. The flux showed only a slight decline in the process with lime/soda ash softening with pH adjustment. Therefore, the experiments with synthetic inorganic water indicate that inorganics cause little membrane fouling during the operational period used in this research, regardless of the various softening processes.

4.3.3 Softening Scenarios with Lake Austin Water

The effect of various softening processes on membrane fouling could be different when natural organic matter is present in the water. Therefore, experiments were performed with Lake Austin water at three lime doses using the variations of the softening process (soda ash, pH adjustment). The operational conditions are shown in Table 4.6. The crossflow velocity was maintained quite constant and the TMPs varied generally within 10 kPa from the designated 90 kPa. The clean water specific fluxes varied from sheet to sheet, with more than a factor of two between the highest and lowest values.

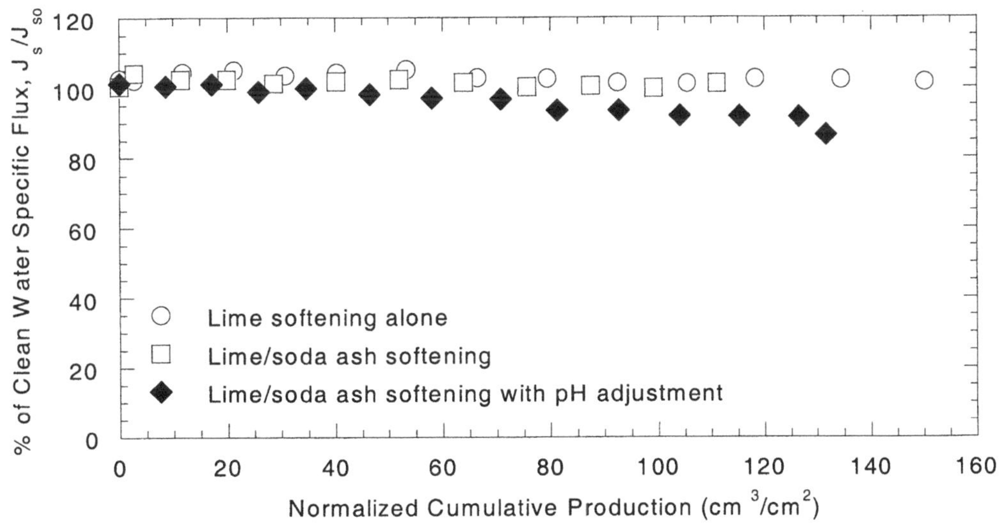

Figure 4.4 Various softening scenarios with inorganic synthetic water at 230 mg/L lime dose

Table 4.6
Operational conditions for each test with the Lake Austin water using the different softening processes.

Lime (mg/L CaO)	Soda ash (mg/L CaO)	pH adjustment	TMP (kPa)	Crossflow velocity (cm/s)	Clean water specific flux, J_{so} (L/m^2-hr-kPa)
125	0	No	90.3-108	10.1	0.881
125	0	Yes	89.0-95.2	10.4	1.997
170	0	No	88.3-100	10.0	1.452
170	0	Yes	94.5-105	10.4	1.685
230	0	No	92.4-97.2	10.0	1.167
230	0	Yes	88.3-100	10.2	1.730
230	105	No	88.3-92.4	10.2	1.731
230	105	Yes	95.2-97.2	10.4	1.730

The water quality achieved at various points in these batch tests is shown in Table 4.7. The results at a 125 mg/L lime dose with and without pH adjustment were quite similar except for the lower pH when re-carbonation was applied. The results of both experiments at the 170 mg/L lime dose were similar to those at a lime dose of 125 mg/L, with some improvement in the TOC, expecially in the experiment with pH adjustment. The results at the high lime dose, i.e., 230 mg/L, reflect reasonable expectations such as high calcium concentrations when soda ash was not used, drastic reduction of Mg^{+2} in all experiments due to the precipitation at high pH, substantial reduction in DOC, and some reduction in SUVA.

The membrane fouling with the Lake Austin water using different softening processes, shown in Figure 4.5, indicated that the flux declined in all experiments, and generally considerably more than in the experiments on the synthetic inorganic water presented above. These results make clear that organic fouling played a bigger role in membrane fouling than inorganic fouling.

At the lime doses of 125 mg/L and 170 mg/L (Figure 4.5, Parts a and b), the pH adjustment had no effect on flux decline. However, the flux decline behavior showed some variations with the lime dose of 230 mg/L (Figure 4.5 c). The greatest flux decline occurred in the experiment with both soda ash addition and pH adjustment; this experiment had an initial flux nearly 100% of the clean water flux, followed by a rapid decline after only 10 cm^3/cm^2 of production, and then a slow decline to approximately 60% of the clean water flux after 120 cm^3/cm^2 of production. The experiment with lime/soda ash without pH adjustment showed the next greatest flux decline, at least as measured from the first to last value in the experiment. The experiment with the least flux decline was that with lime softening alone, although the experiment with lime softening and pH adjustment showed a nearly parallel reduction. The flux with lime softening alone started at 105% and declined only to 90% of the clean water specific flux. The fact that the initial flux was a little greater than 100% of the clean water flux at the experiment could be caused by an operational problem like some air bubbles being trapped in the membrane apparatus during the clean water flux measurement or some difference in temperature between Millipore water and feed water, which was not noted in the measurements.

Table 4.7
Water quality of experiments with the Lake Austin water using the different softening processes

Lime (mg/L CaO)	Soda ash (mg/L CaO)	pH adj.	Samples	pH	Turbidity (NTU)	Ca^{2+} (mg/L)	Mg^{2+} (mg/L)	TOC (mg/L)	SUVA* (L/m-mg)
125	0	No	Raw	8.22	3.3	64.1	20.0	3.6	2.64
			Softened	10.70	27.5	20.3	20.0	2.6	NA
			Feed	9.77	20.3	22.3	20.6	2.8	2.53
			Filtrate	9.74	0.14	14.7	19.4	2.2	2.24
125	0	Yes	Raw	NA	NA	NA	NA	NA	NA
			Softened	10.08	90.0	36.9	18.1	3.6	2.05
			Feed	8.90	47.4	30.3	19.0	3.6	NA
			Filtrate	8.86	0.03	12.1	18.1	3.2	1.85
170	0	No	Raw	8.31	4.6	69.7	21.2	3.3	3.05
			Softened	11.54	5.8	33.5	15.7	2.4	2.94
			Feed	11.25	5.2	27.2	16.0	2.6	2.22
			Filtrate	11.27	0.02	28.3	14.3	2.8	1.38
170	0	Yes	Raw	NA	NA	NA	NA	NA	NA
			Softened	11.11	27.4	32.2	12.7	2.1	NA
			Feed	8.52	0.34	23.2	6.3	1.9	2.13
			Filtrate	8.47	0.11	21.8	6.4	1.7	1.89
230	0	No	Raw	8.10	3.6	NA	19.4	3.7	NA
			Softened	11.49	5.1	76.5	1.2	1.9	
			Feed	11.38	3.6	63.8	0.6	1.9	
			Filtrate	11.40	0.01	64.2	0.5	1.7	
230	0	Yes	Raw	8.29	8.70	74.6	19.1	3.6	2.84
			Softened	11.35	79.0	59.1	6.1	2.5	2.40
			Feed	8.47	22.1	53.5	2.5	1.9	2.20
			Filtrate	8.44	0.13	43.4	2.5	1.7	1.98
230	105	No	Raw	8.12	1.12	59.6	17.3	3.7	2.54
			Softened	11.83	19.6	11.9	3.3	2.4	NA
			Feed	11.70	2.23	4.2	1.4	2.5	1.72
			Filtrate	11.75	0.060	4.2	1.4	2.0	1.71
230	105	Yes	Raw	8.15	2.5	68.5	17.7	4.2	2.3
			Softened	11.98	13.4	15.0	3.0	2.0	NA
			Feed	8.87	2.58	7.9	1.7	2.0	2.1
			Filtrate	8.87	0.013	6.6	1.9	1.6	1.9

* based on TOC, i.e., SUVA=UV254/TOC

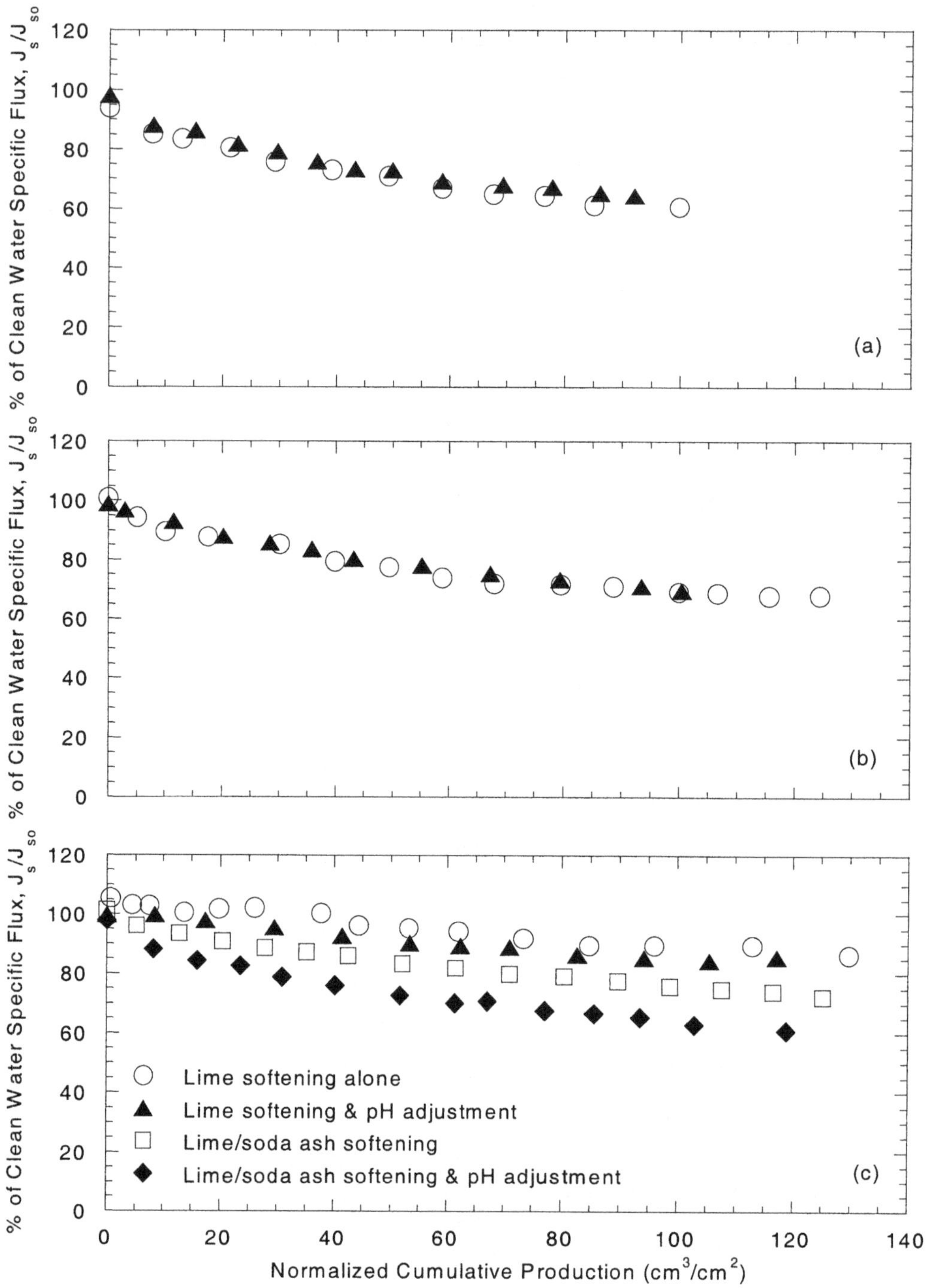

Figure 4.5 Various softening scenarios with Lake Austin water at different lime dose (a) 125 mg/L (b) 170 mg/L and (c) 230 mg/L CaO

The primary result of these experiments is to indicate the importance of natural organic matter in causing membrane fouling. Given the relatively similar flux decline of all four experiments, it is likely that the choice of conditions would be made on other bases than these results. At this high lime dose, where Mg^{+2} precipitation occurs, it is likely that soda ash would be used in real plants. Whether pH adjustment occurred before or after ultrafiltration could be a matter of choice, although most engineers and operators would likely choose to do that before UF to minimize the risk of water quality changes after the membrane (final) treatment.

In most experiments at the enhanced and Mg softening conditions reported in Chapter 5, lime/soda ash softening without pH adjustment was chosen as the standard condition. However, prior to reaching the conclusion that this condition was the best, some experiments had been performed, especially on Lake Austin water, with no soda ash addition. Because the results shown here did not indicate a huge effect of the soda ash addition, it was decided not to perform those experiments a second time. Hence, the experiments at the high lime doses had some variation with respect to soda ash addition.

4.4 SUMMARY

Two natural water sources, Lake Austin water and Missouri River water, were selected for this research and characterized extensively. Experiments at various lime doses were conducted with these two waters to determine the extent of softening. Three softening conditions were chosen corresponding to different removals of hardness ions (Ca^{2+} and Mg^{2+}) and organic matter: standard softening, enhanced softening, and Mg softening conditions. Three lime doses were selected for these three conditions with Lake Austin water, but only two lime doses were selected with Missouri River water because the standard and the enhanced softening conditions occurred at the same dose. The selected lime doses were used throughout this research as posssible pretreatments for the subsequent ultrafiltration.

Several scenarios of softening are used in practice: lime softening alone, lime softening with pH adjustment, lime/soda ash softening, lime/soda ash softening with pH adjustment. The specific process with the least fouling was determined from experiments with inorganic synthetic water and Lake Austin water. Experiments with synthetic inorganic water showed little membrane fouling regardless of the applied softening processes. For Lake Austin water, lime/soda ash softening with pH adjustment showed the greatest flux decline and lime softening alone achieved the least fouling. However, the difference of flux decline in the processes among lime softening alone, lime/soda ash softening, and lime softening with pH adjustment was insignificant. Lime/soda ash softening was chosen to be the standard procedure for this research, but some experiments, particularly with Lake Austin water, were performed with lime softening alone.

One of the methods to describe fouling in ultrafiltration, *i.e.*, flux decline, was normalized to account for variations in pore area and transmembrane pressure between experiments in the laboratory scale system. The flux was normalized with transmembrane pressure and the water flux from the clean membrane. The operational period was normalized with the volume of water treated by a unit area of membrane and the porosity of a standard membrane.

On the basis of the experiments reported in this chapter, several decisions were reached that carried through all of the remaining work. The natural waters and the degree of softening required to reach chosen endpoints for those waters were chosen. The procedures for reporting membrane results in a way that could account for variations among membrane sheets and other operational conditions were also delineated. These methods were used to investigate fouling mechanisms in detail, as reported in Chapter 5.

CHAPTER 5
FOULING IN ULTRAFILTRATION WITH SOFTENING PRETREATMENT

This chapter describes the investigation of four different types of membrane fouling: inorganic fouling, organic fouling, particulate fouling, and combined fouling with organics and particles. Each type of fouling is systematically examined through hydraulic flux decline behavior, effects of three different cleaning methods, and fouled membrane surface analyses with SEM and XPS.

5.1 INORGANIC FOULING

Softening is a process commonly used to remove hardness ions, Ca^{2+} and Mg^{2+}, from drinking water sources. Depending on performance in softening, it seemed possible that the hardness ions remaining in a softened water could cause fouling in ultrafiltration, i.e., inorganic fouling. For instance, the slow kinetics of softening (Nancollas and Reddy 1974, Alexander and McClanahan 1975) could allow precipitation to continue during membrane processes either on the surface or in the pores of membrane. Therefore, inorganic fouling was investigated in this research using synthetic inorganic water, which simulates the inorganic constituents of Lake Austin water without organic matter to avoid effects of organics on softening and ultrafiltration. Three levels of softening were applied, and precipitation kinetics was examined with two shorter flocculation times, 1.5 minutes and 7.5 minutes, than the standard condition (i.e., 30 min of slow mixing).

5.1.1 Extent of Softening for Synthetic Inorganic Water

Three levels of softening (i.e., standard softening, enhanced softening, and Mg softening) represent different degrees of softening with respect to inorganic and organic removal. Considering only inorganic constituents in softening, the standard softening achieves the maximum calcium removal but virtually no magnesium removal; thus calcium carbonate precipitates are dominant in the process. Enhanced softening, which is actually derived from the characteristics of NOM removal, starts to produce some small amount of $Mg(OH)_2$ precipitation. The $Mg(OH)_2$ precipitates become dominant in the Mg softening condition because further $CaCO_3$ precipitation is hindered due to the depletion of carbonate ions, $Mg(OH)_2$ precipitation is almost complete at the high pH (i.e., higher than pH 11), and the $Mg(OH)_2$ precipitates contain a lot more water than the $CaCO_3$ solids.

Table 5.1 summarizes the operational conditions in each test with the synthetic inorganic water at the different lime doses. Transmembrane pressure (TMP) and the crossflow velocity were generally maintained within acceptable ranges, although one experiment had variable pressure. However, the clean water specific fluxes varied considerably among the different membrane sheets.

Table 5.2 shows the water quality at various stages of all of the experiments using the synthetic inorganic water as the raw water. The water quality achieved in each experimental condition was different, as expected. The variation of pH was from 8.47 to 10.39 and turbidity was from 4.2 to 0.5 in the feed water. As expected, the calcium concentration was the lowest at the standard softening condition (125 mg/L CaO) and the magnesium concentration was the lowest at the Mg softening condition (230 mg/L CaO). These batch experiments included a solid/liquid separation after the flocculation (as would be expected in a full-scale operation), so that the solids loading on the membranes (i.e., turbidity in the feed water) was low. As shown below, these water quality differences, including pH, turbidity, calcium concentration, and magnesium concentration, did not seem related to the flux decline behavior in these experiments.

The flux decline in the experiments with synthetic inorganic water at different lime doses using lime softening alone is shown in Figure 5.1. These results are a little scattered, but show that the flux did not decline significantly with any of the three limes doses. The flux at the lime dose of 170 mg/L started at 90% of clean water specific flux and stayed at the same value throughout the entire experiment. The fact that the initial flux was less than 100% of the clean water flux could be caused by operational problems like some air bubbles being trapped in the membrane apparatus, or it could reflect some immediate fouling or blocking membrane pores; however, based on all the membrane results in this research, the latter explanation seems unlikely.

SEM images were taken for the UF sheets after the experiments from the softened inorganic water at the 125 mg/L and 230 mg/L lime doses, as shown in Figure 5.2. The SEM confirmed that a large amount of solids was deposited in some areas, but these solids showed little effect on the flux. The image from the experiment with 125 mg/L of lime (Figure 5.2 a) showed fewer solids on the membrane surface than that with 230 mg/L of lime (Figure 5.2 b), which makes sense because of the higher turbidity in the run with the 230 mg/L lime dose. The dominant shape of deposited solids in both experiments was a well-defined rhombus, which implied that most of them were calcite solids; perhaps, most of the $Mg(OH)_2$ formed in the softening was removed in the settling step.

Table 5.1
Operational conditions of experiments with synthetic inorganic water at different lime doses

Lime dose (mg/L CaO)	TMP (kPa)	Crossflow velocity (cm/s)	Clean water specific flux, J_{so} (L/m^2-hr-kPa)
0	88.3 - 90.3	10.2	1.42
125	76.0 - 106.2*	10.7	1.47
170	94.5 - 99.3	10.0	0.81
230	89.0 - 101.4	10.3	1.90

* unstable TMP

Table 5.2
Water qualities of experiments with synthetic inorganic water at different lime doses

Lime dose (mg/L)	Sample	pH	Turbidity (NTU)	Ca^{2+} (mg/L)	Mg^{2+} (mg/L)
0	Raw	8.46	-	52.7	19.4
	Feed	8.47	-	51.2	19.7
	Filtrate	8.47	-	47.9	19.4
125	Raw	8.16	-	59.3	18.8
	Softened	10.97	11.3	18.4	19.1
	Feed	10.10	0.5	13.9	19.0
	Filtrate	10.02	0.08	13.3	17.8
170	Raw	-	-	-	-
	Softened	11.27	18.2	26.1	6.8
	Feed	10.39	1.5	24.7	2.8
	Filtrate	10.34	0.01	23.8	2.9
230	Raw	8.31	-	36.0	16.5
	Softened	12.02	88.5	33.6	9.5
	Feed	-	4.2	31.9	0.4
	Filtrate	-	0.02	35.7	0.3

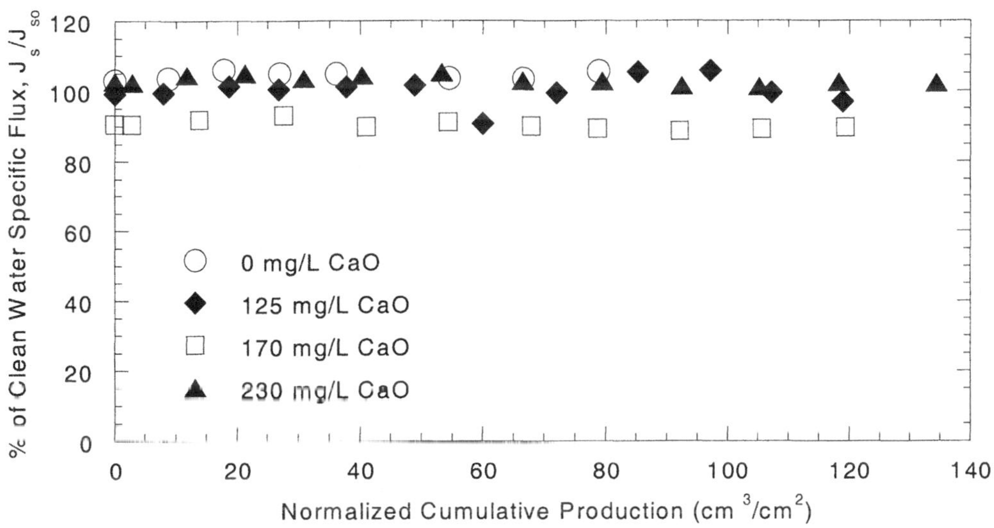

Figure 5.1 Flux decline: effects of softening inorganic water at different lime doses

5.1.2 Precipitation Kinetics

Lime softening has been shown to have slower kinetics than alum coagulation (Nancollas and Reddy 1974, Alexander and McClanahan 1975). Therefore, it is possible that the slow precipitation kinetics in lime softening could affect performance of ultrafiltration after softening. Even after lime softening removes calcium ions, the remaining calcium concentration is relatively high. Therefore, some amount of precipitation might continue to occur during settling and subsequent processes (including membrane processes), unless the precipitation is chemically stopped. The large surface area available in a membrane process can enhance the precipitation of calcium carbonate solids. Theoretically, precipitation could occur either on the surface or in the pores of the membrane, which could cause a flux decline, and therefore needs to be avoided.

The precipitation kinetics was investigated by varying detention times of flocculation at a lime dose of 125 mg/L CaO. In continuous flow experiments, the water flowed directly from a small rapid mix unit (where lime was added) through a flocculation reactor and then to the ultrafiltration (UF) unit without settling. The detention time in the flocculation reactor was adjusted in different experiments using different volumes of the reactor (since the flow rate was determined by the needs of the membrane system). The standard condition in the batch softening jar tests used 30 minutes of detention time. In the continuous flow tests, the detention time in flocculation was reduced to 7.5 minutes and 1.5 minutes. Continuous flow experiments allow the filtration rate to be matched by a sample feed line from the flocculation tank, and therefore obtain steady state conditions with respect to softening. Synthetic inorganic water that simulated the inorganic constituents in Lake Austin water was used for the experiments.

The operational conditions of these tests are summarized in Table 5.3. The continuous flow experiments were limited to a low crossflow velocity, since the short flocculation time was impossible to obtain with recycling and the UF filtration rate had to match the feed flow from the earlier units. The TMP values were very unstable due to the low crossflow velocity and the large amounts of solids, which were unsettled flocs.

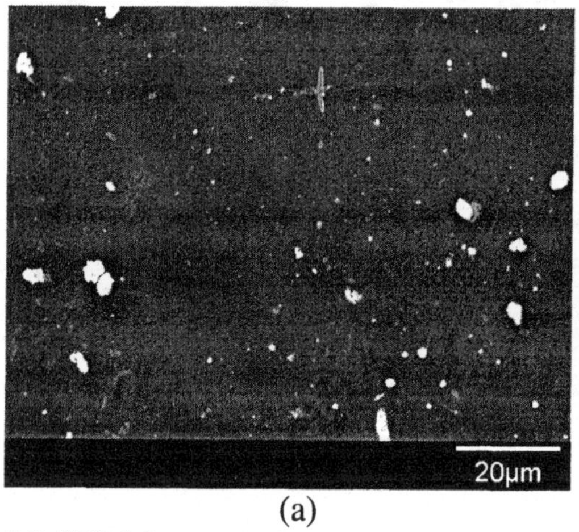

(a)

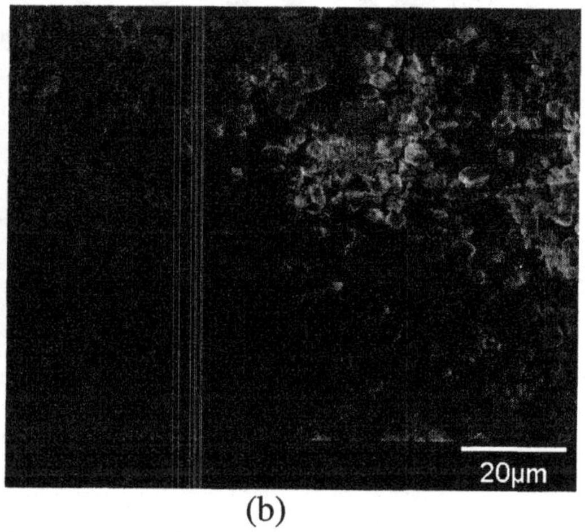

(b)

Figure 5.2 SEM images of membranes fouled by inorganic water softened at (a) 125 mg/L and (b) 230 mg/L CaO

Table 5.3
Operational conditions of experiments for precipitation kinetics (lime dose: 125 mg/L CaO)

Precipitation time (min)	TMP (kPa)	Crossflow velocity (cm/s)	Clean water specific flux, J_{so} (L/m^2-hr-kPa)
1.5	95.9 - 208.3*	1.7	0.71
7.5	91.7 - 193.8*	1.7	1.70

* very unstable TMP

The water quality at various stages was measured for the experiments with precipitation time of 1.5 minutes. The rise in pH (i.e., pH 11.16) was certainly due to the lime addition and the pH was higher than what was expected (i.e., between approximately pH 10.5 and 11). The turbidity increased considerably in the feed water (i.e., 449 NTU) in the continuous flow tests, reflecting the precipitation of the calcium carbonate. The total calcium increased between the raw water (47 mg/L as Ca^{2+}) and feed water (120.4 mg/L as Ca^{2+}) because of the lime addition, but the soluble calcium (12.7 mg/L as Ca^{2+}) was similar to the filtrate values (11.1 mg/L as Ca^{2+}). At this low dose of lime, magnesium drops only a small amount from 18.6 to 16.1 mg/L as Mg^{2+}, presumably due to co-precipitation with the calcium (i.e., incorporation of some magnesium into the calcium carbonate solids formed).

As shown in Figure 5.3, the flux decline was not substantial for the waters softened at 125 mg/L of lime using 1.5 and 7.5 minutes of precipitation times. The scatter in the experiment with 7.5 minutes of precipitation stems from the combination of high solids content and low crossflow velocity. Despite the scatter, the results are sufficiently clear to see that the flux decline was quite small, a surprising result given the high turbidity (449 NTU). The small flux decline in both experiments indicated that the pores were not blocked by precipitation within them. It had been hypothesized that precipitation could occur in the membrane system by the slow precipitation kinetics and, if it occurred inside the pores, would severely foul the membrane. However, it seems that precipitation either was completed prior to reaching the membrane or occurred entirely on the external surface, not inside the pores. The SEM images from the experiment in Figure 5.4 show well the precipitates on the membrane surface. The petal-shaped image in the left portion of the figure might be interpreted as precipitation that occurred on the membrane surface.

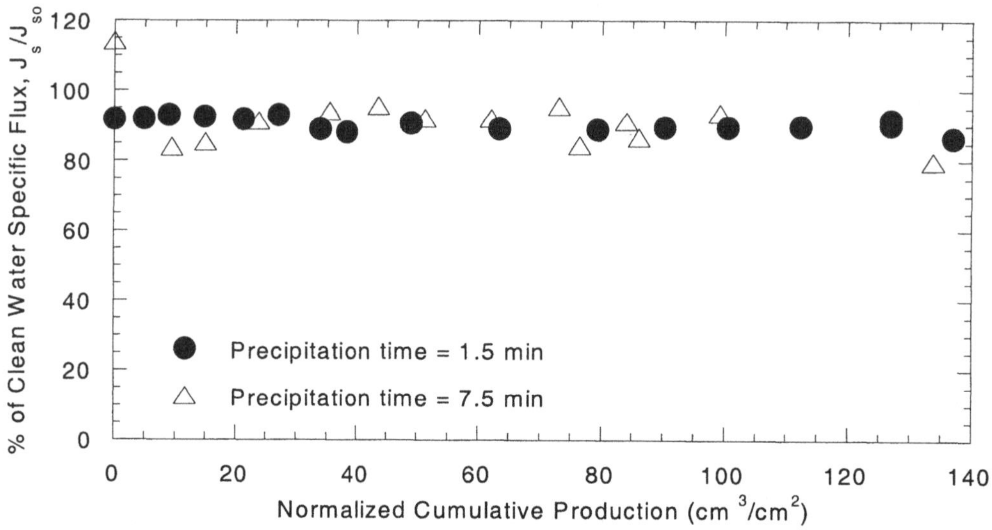

Figure 5.3 Flux decline: effects of different precipitation times using the synthetic inorganic water (125 mg/L CaO)

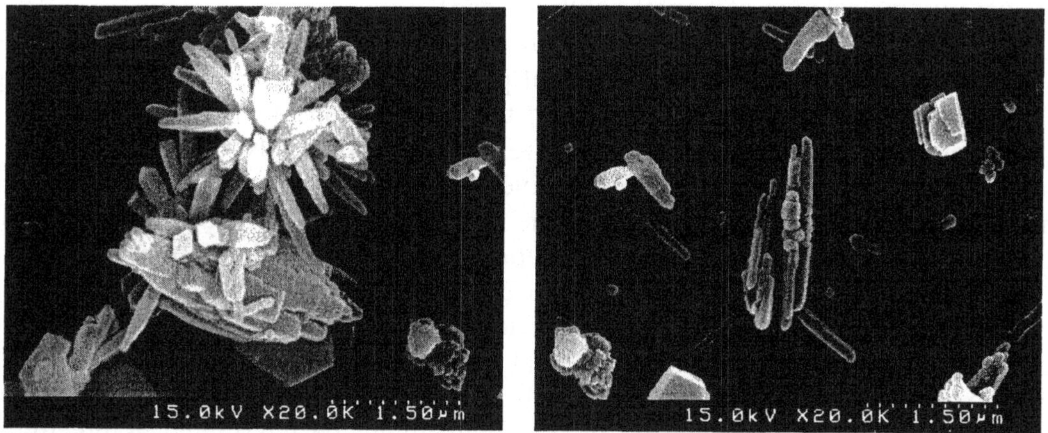

(a) (b)

Figure 5.4 SEM images of membranes fouled by the inorganic water softened at 125 mg/L CaO (precipitation time: 7.5 min)

5.1.3 Summary of Inorganic Fouling

The inorganic fouling in ultrafiltration after softening pretreatment was investigated with synthetic inorganic water. Softening was performed at different levels of lime doses, which corresponded with different Ca^{2+} and Mg^{2+} concentrations and dominant precipitates. The results showed little effect on water flux decline in ultrafiltration regardless of the extent of softening. The SEM images from the membrane at the lime dose of 230 mg/L CaO made it clear that a great amount of deposits on the membrane surface had virtually no effect on water flux.

In addition, the possible effects in membrane fouling from the slow precipitation kinetics in softening were investigated using short precipitation times such as 7.5 and 1.5 minutes.

Synthetic inorganic water was used to avoid any NOM effect. The flux decline results showed that the slow kinetics had little significance in terms of inorganic fouling.

5.2 ORGANIC FOULING BY SIMPLE ORGANIC COMPONENTS

Natural organic matter (NOM) is one of the major foulants in membrane processes. Therefore, much research has been performed to understand its role in fouling, including various solution characteristics and effects of different functional groups (Mallevialle, Anselme, and Marsigny 1989; Laine et al. 1989; Jucker and Clark 1994; Chellam et al. 1997; Hong and Elimelech 1997).

However, NOM has a variety of functional groups and complicated structures, so it is difficult to obtain a complete understanding. Many attempts have been made to separate NOM into different fractions: fractionation by molecular weights (Lin, Lin, and Hao 2000), fractionation by hydrophobicity using XAD-8 resin (Jucker and Clark 1994, Nilson and DiGiano 1996, Caroll et al. 2000), and fractionation by substructure using GC-pyrolysis-MS (Mallevialle, Anselme, and Marsigny 1989, Lahoussine-Turcaud et al. 1990, Mackey and Wiesner 1999).

To overcome the complexities in interpreting NOM effects on membrane fouling, simple organic components were considered as surrogates for NOM in this research. Previous research on NOM components in membrane processes indicated that polysaccharides formed an important part of the macromolecular solutes present in the fouling cake (Mallevialle, Anselme, and Marsigny 1989, Lahoussine-Turcaud et al. 1990, Mackey 2000). Therefore, several polysaccharides were evaluated in this research for their removal in softening, and the selected polysaccharides were further investigated for their effects on membrane fouling at various softening conditions.

5.2.1 Primary Screening for NOM Surrogates

A simple organic component, i.e., polysaccharide, was proposed to use as a NOM surrogate based on previous research. As a preliminary test to choose two compounds for further study, the removal efficiency of three polysaccharides (dextran, alginic acid, and polygalacturonic acid) and a smaller carbohydrate (tannic acid) in the softening process was evaluated. Each organic component was tested at two concentrations, 4 and 20 mg/L as carbon.

For each concentration of each chemical, six jars were softened under different conditions. Each jar contained one liter of synthetic organic water, which was made from the specific organic chemical and the same inorganic constituents as the Lake Austin water. The first jar was used as a control, meaning that no lime was added but the same softening procedures as other jars were followed. The next three jars used the three standard lime doses, i.e., 125, 170, and 230 mg/L, with no soda ash addition. The last two jars were run with soda ash addition for the two highest lime doses: 45 mg/L as CaO of soda ash (i.e., 85.1 mg/L as Na_2CO_3) was added for the 170 mg/L lime dose and 105 mg/L as CaO of soda ash (i.e., 198.5 mg/L as Na_2CO_3) was added for 230 mg/L lime dose. The soda ash doses were based on the result that the calcium concentration was the minimum at 125 mg/L lime dose and on the 1:1 stoichiometry between calcium and carbonate.

The screening results from the 4 mg/L of dextran solution are shown in Figure 5.5; several parameters related to softening and organic matter removal within softening are shown in different parts of the figures. Dextran is available in several molecular weights, and the results reported here were done with dextran with a molecular weight of 12 kDa. The results are shown along with the data for the softening of Lake Austin water by lime alone, because the goal of these experiments was to choose organic components that had removal behavior similar to NOM. The pH rise (Figure 5.5 a) with increasing lime dose for the dextran solution was similar to that with the Lake Austin water, although the pH was slightly higher for the synthetic water. The calcium (Figure 5.5 b) and magnesium (Figure 5.5 c) removals of the dextran solution when no soda ash was added were also similar (but not identical) to those of Lake Austin water. The Ca^{2+} removal of the dextran solution was very close to that of Lake Austin water to the point with the minimum calcium concentration, but at higher lime doses, the calcium concentration in the dextran solution was higher than in the natural water. These results make sense because Lake Austin water has more alkalinity from other low concentration species such as silicate. Magnesium also was considerably less in the dextran solution than the natural water at the two higher doses, probably because of the higher pH. The precipitation of $Mg(OH)_2$, of course, is highly sensitive to pH in this range. With the soda ash addition, the calcium and magnesium removals were dramatic, with concentrations near zero at the 230 mg/L lime dose.

The most important result was that the DOC removal (Figure 5.5 d) was almost the same as Lake Austin water in the solution with 4 mg/L dextran. The UV absorbance at 254 nm and SUVA are not shown since dextran does not absorb UV light at 254 nm, so that measure and the related specific UV absorbance (SUVA) were extremely low. These surrogate measures for NOM are inappropriate as a measure of dextran behavior.

The screening results from 20 mg/L C of dextran solution had similar patterns to the results from 4 mg/L; the pH rise close to Lake Austin water, the Ca^{2+} removal close to that of Lake Austin water to the point with the minimum but a higher calcium concentration at higher lime doses, and the low Mg concentration at higher doses due to higher pH. The results from alginic acid at 4 mg/L and 20 mg/L C and polygalacturonic acid at 4 mg/L C also showed that the inorganic measures (pH, Ca^{2+}, and Mg^{2+}) were similar to the dextran solutions (and to Lake Austin water) as lime doses were increased. Therefore, the screening results of inorganic measures (pH, Ca^{2+}, and Mg^{2+}) from other components such as alginic acid and polygalacturonic acid are not shown.

In addition, we attempted to investigate tannic acid with the same procedure as the other organics. However, in trying to make a 4 mg/L solution in the synthetic organic water from a concentrated tannic acid solution (in distilled deionized water), the solution became quite turbid, indicating that a precipitate was formed. After filtration through a 0.45 μm membrane, the DOC was measured as 0.8 mg/L C; the fact that this measure was far less than the 4 mg/L target confirmed the precipitation. Hence, tannic acid apparently forms a precipitate with the calcium or magnesium in a hard water at a pH of approximately 8.3, and is therefore an inappropriate surrogate for NOM in this research.

The most important result from the screening tests is the DOC removal; to be used as an NOM surrogate, the specific organic must give a similar removal as was found for DOC in Lake Austin water. The DOC removals of dextran and alginic acid at 4 and 20 mg/L C are shown together in Table 5.4. The DOC removal of dextran at 4 mg/L C was almost the same as Lake Austin water. For the dextran solution at 20 mg/L C, the DOC was gradually removed but the removal was slightly lower than the Lake Austin water on a percent basis. The DOC removal of alginic acid was more dramatic than NOM (and dextran) with increasing lime doses, but nevertheless showed a real trend of increasing removal with increasing dose like NOM in Lake Austin.

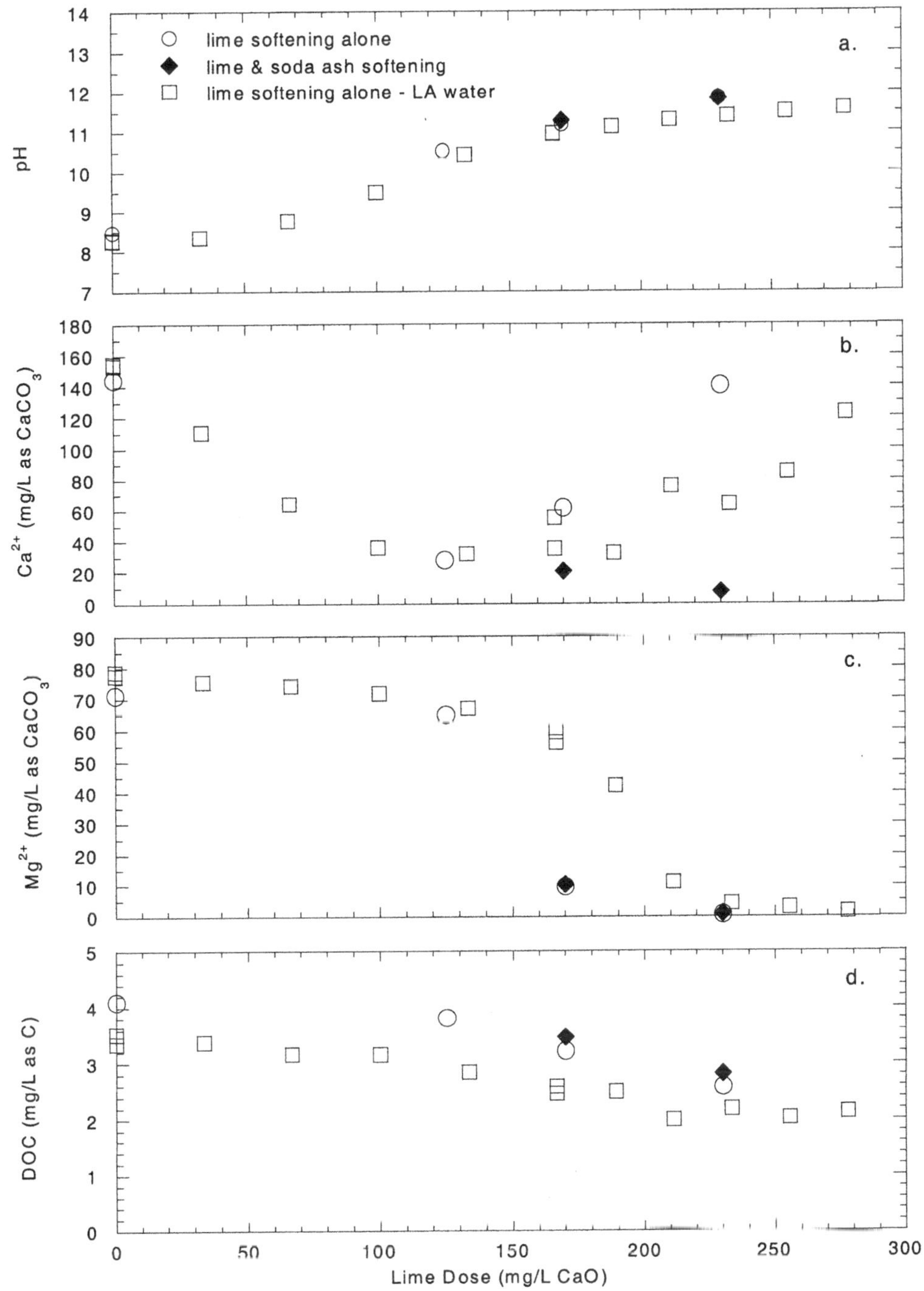

Figure 5.5 Primary screening results of 4 mg/L C of dextran solution

The results from polygalacturonic acid were obtained from just 4 mg/L C of the solution (not shown). The results revealed that the DOC removal was almost complete at the 125 mg/L lime dose, and no further DOC removal occurred with increasing lime doses. Therefore, the polygalacturonic acid would not be useful as a surrogate for NOM in this research because the degree of softening made no difference in the DOC removal.

Based on the screening test, the NOM surrogates could be dextran and alginic acid since their behavior in softening was more analogous to Lake Austin water than the polygalacturonic acid and tannic acid solutions.

5.2.2 Investigation of Dextran on Membrane Fouling

Effects of Nominal Molecular Weights of Dextran

Dextran has been extensively used to determine nominal molecular weight cut-offs (MWCO) of membranes. Its dense properties prevent deformation of molecules under transmembrane pressures commonly experienced during membrane operations and its relatively round shape helps to determine pore sizes. In this research, three nominal molecular weights (12, 60, and 500 kDa) and two concentrations (4 and 20 mg/L C) of dextran were used to investigate their effects on ultrafiltration. To differentiate the dextran solutions with the different nominal molecular weights, dextran was designated as Dextran12, Dextran60, and Dextran500, where the numbers refer to the molecular weights in kDa.

Table 5.4
DOC results from the screening tests of dextran and alginic acid at two concentration (4 and 20 mg/L C)

Samples		Lime Doses (mg/L CaO)			
		0	125	170	230
Lake Austin	Conc. (mg/L)	3.4	2.9	2.5	2.0
	% Removal	-	14.7	26.5	41.2
Dextran – 4 mg/L	Conc. (mg/L)	4.1	3.8	3.2	2.6
	% Removal	-	7.3	22.0	36.6
Dextran –20 mg/L	Conc. (mg/L)	20.7	19.5	18.8	17.1
	% Removal	-	5.8	9.2	17.4
Alginic acid – 4 mg/L	Conc. (mg/L)	4.6	2.2	0.9	0.1
	% Removal	-	52.2	80.4	97.8
Alginic acid – 20 mg/L	Conc. (mg/L)	16.9	3.7	0.7	0.2
	% Removal	-	78.1	95.9	98.8

Table 5.5
Operational conditions of experiments with raw dextran (without softening)

Nominal molecular weight (kDa)	DOC (mg/L C)	TMP (kPa)	Crossflow velocity (cm/s)	Clean water specific flux, J_{so} (L/m^2-hr-kPa)
12	4	88.3 - 95.2	10.5	1.89
12	20	86.9 - 92.4	9.6	1.67
60	4	90.3 - 95.2	10.1	1.95
60	20	88.3 - 91.0	10.2	1.56
500	4	86.2 - 91.0	9.7	1.91
500	20	88.3 - 89.7	10.3	1.87

Table 5.5 summarizes operational conditions in each test with raw dextran solutions, *i.e.*, dextran without any softening treatment. Transmembrane pressure (TMP) fluctuated slightly (between 88.3 and 95.2 kPa) around the target value of 90 kPa. The crossflow velocity was sustained at approximately 10 cm/s. The variation in the clean water specific flux among the membrane sheets in these experiments was fairly small, with all the values in the range of 1.6 to 1.9 L/m^2-hr-kPa.

Table 5.6 shows the water quality of all experiments using the raw dextran solution as the feed water. The water quality of inorganic constituents such as Ca^{2+} and Mg^{2+} in each experiment was essentially unchanged because no softening process was applied.

The essential result of these experiments is that the nominal molecular weight of dextran has a dramatic effect on DOC removal by ultrafiltration. The removal of Dextran12 with the membrane used in the experiment was marginal, just 2% when 20 mg/L of dextran was treated. One might expect a high percent removal because the reported molecular weight cut-off of membranes (10 kDa) is less than the reported molecular weight of the dextran (12 kDa). Apparently, either the MWCO was higher than the reported value or much of the dextran had a lower MW than its reported value, or both. In any case, virtually all of the dextran could pass the membrane.

Table 5.6
Water quality of experiments with raw dextran (without softening)

Nominal molecular weight (kDa)	DOC (mg/L C)	Sample	pH	Ca^{2+} (mg/L)	Mg^{2+} (mg/L)	DOC (mg/L)
12	4	Feed Filtrate	NA	NA	NA	NA
12	20	Feed Filtrate	8.61 8.58	63.0 61.0	16.8 17.1	19.4 19.0
60	4	Feed Filtrate	8.26 8.25	48.2 45.0	16.9 15.6	5.4 2.7
60	20	Feed Filtrate	8.41 8.45	59.6 59.8	17.9 17.6	22.0 19.4
500	4	Feed Filtrate	NA	NA	NA	4.6 0.0
500	20	Feed Filtrate	8.51 8.55	56.1 57.8	16.4 16.6	24.9 0.3

With the Dextran60, the DOC removals increased to approximately 50% at 4 mg/L C. Some molecules smaller than the nominal molecular weight could be present and pass the membrane (an explanation confirmed with molecular weight measurements by size exclusive chromatography shown subsequently). However, at the higher concentration, i.e., 20 mg/L, the DOC removal was reduced to approximately 11%. The rejected dextran from the solution at the high concentration accumulated on the membrane surface, and thus might have had more chance to pass through the membrane.

The Dextran500 was completely removed by the membrane (almost 100%), regardless of the DOC concentrations of 4 or 20 mg/L C. The absolute removal of the Dextran500 was expected because the molecular size, 500 kDa, is far greater than the MWCO of the membrane.

The reductions in flux in the experiments with raw dextran solution with different nominal molecular weights are shown in Figure 5.6. Figure 5.6 (a), with the results from the 4 mg/L C experiments, clearly demonstrates effects of molecular sizes of dextran solutions on membrane fouling at the relatively low DOC concentration. The smallest dextran (Dextran12), similar to the MWCO of membrane, showed no flux decline; the medium size dextran (Dextran60) had a flux decline to 60% of the clean water specific flux; and the largest dextran (Dextran500) showed the biggest flux decline to 40% of the clean water specific flux. These results illustrate that physical sieving is one of the main removal mechanisms in membrane processes, as expected. However, with the sieving mechanism alone, one cannot explain the relative flux decline among the three different molecular sizes of dextran.

Figure 5.6 (b) reveals that effects of a high concentration of dextran could diminish the effects of molecular sizes on membrane fouling. The smallest dextran, i.e., Dextran12, showed no flux decline even at the high dextran concentration (i.e., 20 mg/L C of DOC) because there was so little removal. However, experiments with Dextran60 and Dextran500 showed almost identical flux decline despite the differences in molecular sizes and the dramatic difference in removal. This result might suggest that only certain amounts of foulant deposited on membrane are critical to membrane fouling.

Effects of Extent of Softening on Fouling by Dextran

The dextran with the nominal molecular weight of 60 kDa (Dextran60) was selected to investigate further the effects of the degree of softening pretreatment prior to ultrafiltration because only the Dextran60 showed some differences in membrane performance with the concentrations of our interest (i.e., 4 and 20 mg/L). The raw dextran solution was softened at the same three levels of softening based on the softening performance in the Lake Austin water used throughout this research. As before, soda ash was added in some experiments at the high lime doses, i.e., 170 and 230 mg/L CaO to facilitate continuous $CaCO_3$ precipitation by supplying CO_3^{2-} ions.

Table 5.7 summarizes operational conditions in each test with the Dextran60 solution at different lime doses and two concentrations. The crossflow velocities and TMPs were maintained in the proper ranges during the operation of each experiment. The clean water specific fluxes fluctuated between different experiments. It seemed that the membranes came from two different batches, since three experiments had clean water specific fluxes of approximately 2.0 L/m^2-hr-kPa and the other three had values of approximately 1.6 L/m^2-hr-kPa.

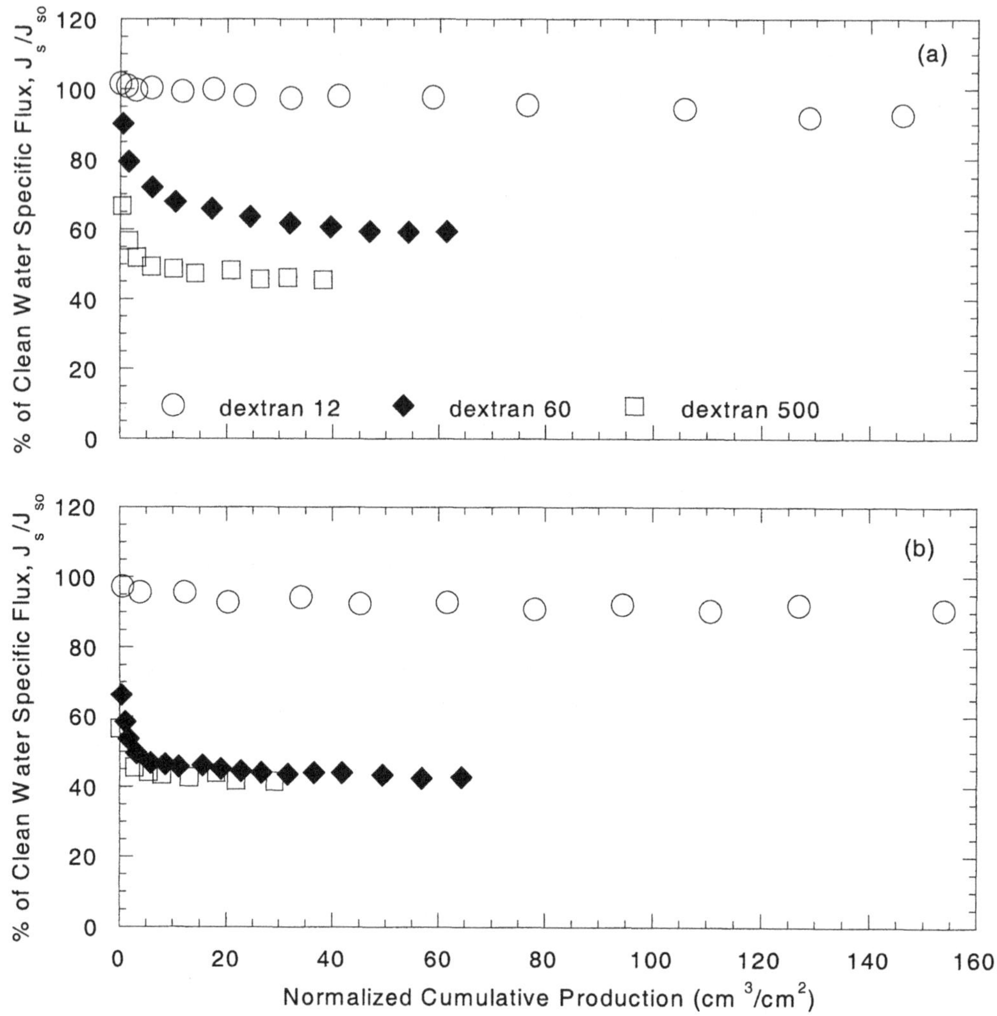

Figure 5.6 Flux decline: effects of nominal molecular weights of raw dextran at two DOC concentrations: (a) 4 mg/L and (b) 20 mg/L

The water quality in the experiments with the Dextran 60 at different lime doses and two concentrations is presented in Table 5.8. The water quality achieved in each experimental condition was different because of the different degrees of softening (and consequent different degrees of removal of the organic matter). For instance, the variation of pH was from 8.25 to 11.37 and the calcium concentration was from 59.2 to 4.4 mg/L as Ca^{2+}in the water used as feed to the membrane. The pH values in the experiment with 4 mg/L C of Dextran60 solution softened at 170 mg/L of lime are shown as lower than those with softened at 125 mg/L of lime; in the experiment at 170 mg/L of lime, the pH meter was erroneously calibrated so that the pH meter was reading low by approximately 0.3 – 0.4 unit.

Table 5.7
Operational conditions of experiments with Dextran60 at different degrees of softening and two concentrations (4 and 20 mg/L C)

DOC (mg/L C)	Lime (mg/L CaO)	Soda ash (mg/L as CaO)	TMP (kPa)	Crossflow velocity (cm/s)	Clean water specific flux, J_{so} (L/m^2-hr-kPa)
4	0	0	90.3 - 95.2	10.1	1.95
4	125	0	89.7 - 93.1	10.0	2.16
4	170	45	91.7 - 95.9	10.1	1.93
4	230	105	88.3 - 92.4	9.7	1.64
20	0	0	88.3 - 91.0	10.2	1.56
20	170	45	91.0 - 96.6	9.8	1.56

Table 5.8
Water quality of experiments with Dextran60 at different degree of softening and two concentrations (4 and 20 mg/L)

Designated DOC (mg/L C)	Lime (mg/L CaO)	Soda ash (mg/L CaO)	Sample	pH	Turbidity (NTU)	Ca^{2+} (mg/L)	Mg^{2+} (mg/L)	DOC (mg/L)
4	0	0	Feed	8.26	NA	59.2	16.9	5.4
			Filtrate	8.25		45.0	15.6	2.7
4	125	0	Raw	8.29	NA	50.8	17.0	5.8
			Softened	10.83	47.1	19.6	13.9	4.5
			Feed	10.56	2.03	12.8	13.6	4.1
			Filtrate	10.52	0.014	14.0	13.6	2.7
4	170	45	Raw	8.06*	NA	56.9	16.6	4.6
			Softened	10.77*	27.2	20.5	3.0	1.9
			Feed	10.41*	4.20	11.1	2.2	1.8
			Filtrate	10.55*	0.013	7.8	1.9	0.7
4	230	105	Raw	8.45	NA	53.8	18.3	4.4
			Softened	11.52	6.8	7.2	0.8	1.3
			Feed	11.37	3.4	4.4	0.6	1.0
			Filtrate	11.42	0.011	4.4	0.5	0.3
20	0	0	Feed	8.41	NA	59.6	17.9	22.0
			Filtrate	8.45		59.8	17.6	19.4
20	170	45	Raw	8.45	NA	63.2	19.3	21.7
			Softened	11.03	21.7	9.1	4.8	17.2
			Feed	10.69	2.70	7.1	3.7	16.6
			Filtrate	10.72	0.014	8.7	3.6	14.2

* Erroneous calibration of pH meter: low by approximately 0.3-0.4 unit

In addition, the experiments with the low DOC concentration showed decreases of DOC in the feed water from 5.4 to 1.0 mg/L C as the lime doses increased from 0 mg/L to 230 mg/L CaO; DOC removal by softening was expected from the screening test. However, DOC removal rates showed some discrepancies between the results from the screening test and these experiments. The DOC removal rates were 7% at 125 mg/L CaO and 37% at 230 mg/L CaO in the screening tests, which were performed with the lower (12 kDa) molecular weight dextran, i.e., Dextran12. During these experiments using the Dextran60, the DOC removal rates were 23% at 125 mg/L CaO and 71% at 230 mg/L CaO. These results suggest that the molecular size of organic matter might have a role in organic removal mechanisms in softening such as co-precipitation and surface adsorption. The experiments with 20 mg/L of DOC concentration showed a similar trend of DOC removal. At 170 mg/L of lime, the DOC removal was 9% in the screening test and 20% in these experiments. Therefore, the results illustrate that DOC removal by softening increased somewhat as the nominal molecular weight of dextran increased from 12 kDa to 60 kDa. The DOC removal by ultrafiltration was in the range of 33% to 66% of the feed water, which is expected since the size of dextran (Dextran60) is relatively big compared to the membrane nominal pore size (10 kDa).

The flux decline in the experiments with 4 mg/L of the Dextran60 at different lime doses is shown in Figure 5.7. The results show that removal of the dextran by softening reduces membrane fouling. After softening the dextran solution at 125 mg/L of lime, the flux leveled off at approximately 73% of the clean water specific flux at approximately 60 cm^3/cm^2 of the normalized cumulative production, while the flux was 59% of the clean water specific flux without softening after the same production. As the degree of softening was increased to the "enhanced softening" condition with 170 mg/L of lime, virtually no flux decline was observed.

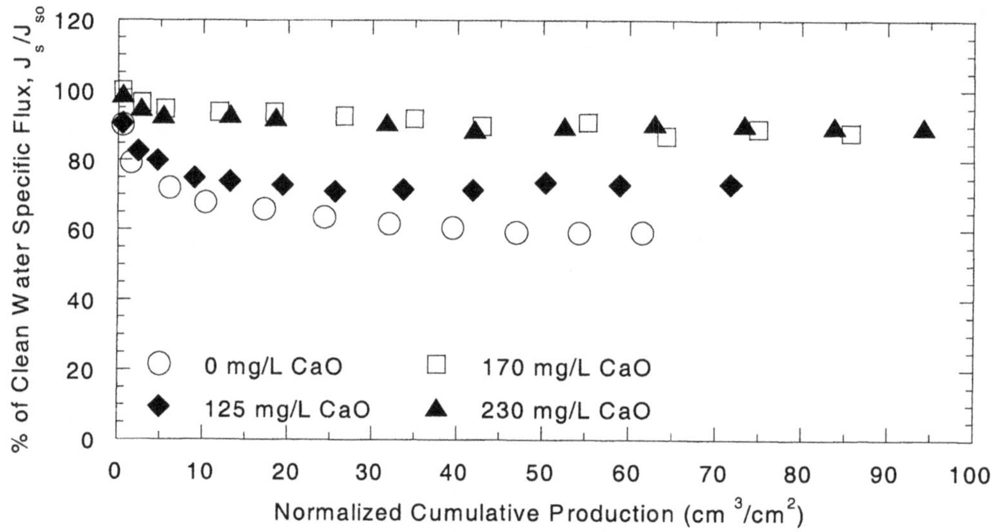

Figure 5.7 Flux decline: effects of softening Dextran60 at different lime doses (DOC: 4 mg/L C)

Note that the DOC concentration in feed water was 4.1 mg/L at the lime dose of 125 mg/L CaO and 1.8 mg/L at the lime dose of 170 mg/L CaO as shown in Table 5.8. The results imply that the portion of dextran solution that is primarily responsible for fouling in membranes was preferentially removed by softening. In addition, the reduced concentration of the doubly-charged calcium and magnesium ions, which perhaps help attach organic matter to a membrane surface, could have aided the reduction in the flux decline.

Effects of Softening on Fouling at Different Concentrations of Dextran

The results from the experiments with softening at different lime doses show that softening pretreatment can substantially improve the water flux under the relatively low dextran concentration (4 mg/L C). Softening was applied to waters with a high concentration of dextran (20 mg/L C) to better understand its effect on fouling.

The operational conditions and water quality are presented in Table 5.7 and Table 5.8 in the previous section. Flux decline patterns from the experiments with two concentrations (4 and 20 mg/L C) are shown in Figure 5.8. Softening was performed at the lime dose 170 mg/L CaO. The result was compared with the raw waters. At the low dextran concentration, softening showed dramatic effects on the reduction of flux decline, i.e., 90% of clean water specific flux by softening at 170 mg/L CaO but 60% of clean water specific flux with the raw dextran solution after approximately 60 cm^3/cm^2 of the normalized cumulative production. At the high concentration (20 mg/L C), the flux improved to 54% of the clean water specific flux with the enhanced softening compared to 43% of the clean water specific flux with the raw Dextran60 solution at the same production. At the lime dose of 170 mg/L CaO, the flux improvement in the experiment with 20 mg/L C of Dextran60 was not as much as with 4 mg/L of Dextran60, where very little flux decline was shown. The DOC results with experiments at the lime dose of 170 mg/L CaO help explain the difference; the feed water still had 77% of the original DOC at the high concentration, whereas only 23% of the DOC remained at the low concentration of dextran (as shown in Table 5.8). These results suggest that the capacity for DOC removal in softening might be limited by the amount of precipitation, i.e., by adsorption capacity.

Efficiency of Cleaning Methods on Dextran

In addition to the improvement of hydraulic water flux, another possible advantage of pretreatment for a membrane process is to condition fouling cakes to make them more easily recovered by cleaning procedures. Three cleaning methods were used sequentially in this research: a surface wash with distilled/deionized water, a caustic wash with a high pH solution (*i.e.*, 0.5N NaOH), and an acidic wash with a low pH solution (*i.e.*, 0.1 N HNO_3). A low pH solution is used to remove inorganic precipitates, and a high pH solution is more effective to remove organic foulants.

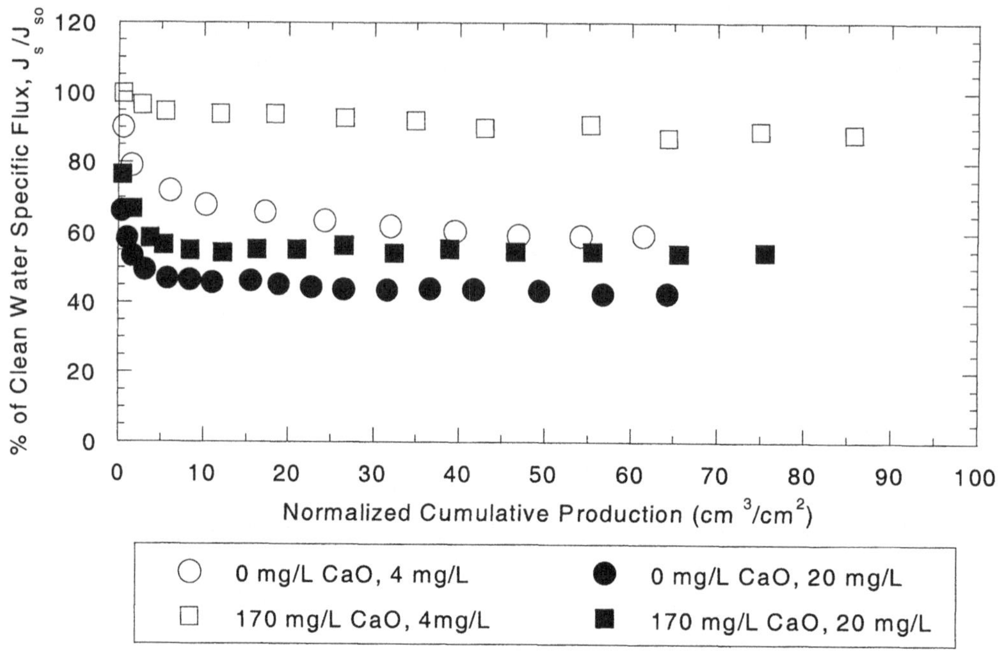

Figure 5.8 Flux decline: effects of softening Dextran 60 at different concentrations

After each cleaning, the clean water specific flux was measured using distilled/deionized water for 20 minutes. These measurements after each cleaning method were to evaluate the efficiencies of cleaning processes and therefore determine the mechanisms of fouling. The results of the efficiency of each cleaning for membranes fouled with dextran are shown in Figure 5.9. After each cleaning, the flux was measured at three times (5, 10, and 20 minutes), and, at each time, three separate measurements were performed. The means and standard deviations of the nine measurements are shown in the figure. The results show a very narrow range of the standard deviations of flux measurements, *i.e.*, between 1.3% and 4.3%, which indicates well controlled flux measurements. Similar low standard deviations for the flux measurements were found in all experiments reported herein.

Only the experiments with lime doses at 125 mg/L and 170 mg/L were evaluated along with the raw Dextran60 solution because the flux decline at 230 mg/L of lime was almost the same as that at 170 mg/L of lime. The flux at the end of run with dextran at 170 mg/L of lime was 88% of the clean water specific flux; thus the efficiency of the cleaning methods was hard to evaluate properly with just 12% of the initial flux possible to recover. As shown in Figure 5.9, surface wash yielded little or no improvement in the flux, suggesting either that fouling occurred inside of the pores or the fouling cake was attached chemically to the membrane surface. After the caustic wash, 65% of the lost flux was recovered for the raw Dextran60 solution, while 46% of the lost flux was recovered for the solution treated at 125 mg/L of lime. The caustic wash, which removes organic matter in the fouling layer, was quite effective in cleaning the membranes fouled with the dextran solutions. Also, the acidic wash recovered an additional 5 to 10% of the clean water specific flux. Taken together, these washes yield a recovery of about 93% of the initial clean water specific flux.

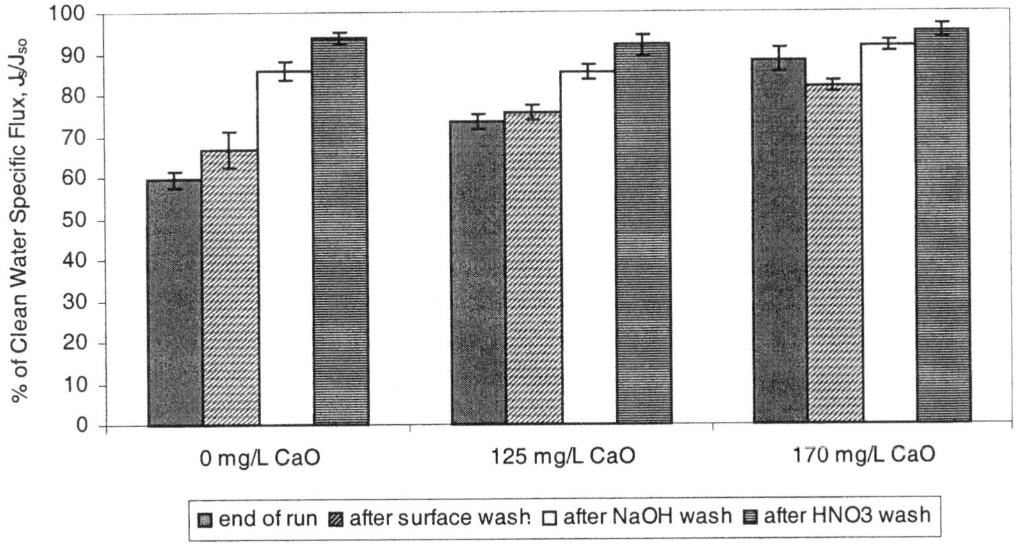

Figure 5.9 Membrane cleaning: effects of softening Dextran60 at different lime doses (4 mg/L C)

Molecular Weight Distribution in Softening of Dextran

Previous research indicated that organic matter with high molecular weights has been preferentially removed in water treatment processes such as softening and coagulation (Liao and Randtke 1986, Semmens and Staples 1986, Tambo and Kamei 1989). Especially Liao and Rantdke (1986) implied that excellent removal of humic substances and proteins could be achieved in softening because of substantial dissociation of functional groups under the high pH conditions.

The molecular weight distributions of the raw and softened Dextran60 solutions at three lime doses were analyzed with size exclusion chromatography (SEC) as presented in Figure 5.10. The abscissa, the elution volume, is inversely proportional to molecular weight since the higher molecular weights (MW) of organic matter pass through column faster than the lower MW molecules. The calibration was performed with four dextran standard solutions: 1.08, 10.5, 66.7, and 401.3 kDa (Fluka, St. Louis, MO). Then, a relationship between a molecular weight and elution volume/mL was calculated. The ordinate (*i.e.*, height of peak) is proportional to the concentration. Each sample originally had a different peak height because its concentration varied after softening and the difference was exaggerated after being concentrated by the rotary evaporator. Since the purpose of analyzing the molecular weight distribution was to learn whether softening selectively removes high molecular weights of dextran, the ordinate was fitted to a common scale (as shown in Figure 5.10).

The Figure 5.10 (a) shows the molecular weight distribution of the raw Dextran60 with the median value of molecular weights. The average molecular weight of Dextran60 reported by manufacturer is 60 kDa, which is close to the median value found in these SEC measurements (*i.e.*, 55.5 kDa). In general, the SEC analyzes three different average molecular weights: the number-average molecular weight and the weight-average molecular weight. With these average molecular weights, the degree of polydispersity (PD) can be obtained as follows:

$$PD = \frac{MW_w \text{ (the weight - average molecular weight)}}{MW_n \text{ (the number - average molecular weight)}} \quad (5.1)$$

The degree of polydispersity indicates the shape of molecular weight distribution, with a high value of PD indicating a wide distribution of molecular weight. The result of the raw Dextran60 shows a narrow distribution of molecular weight since the polydispersity degree was 1.9 (*i.e.*, 79.5 kDa for MW_W and 41.7 kDa for MW_n). Other samples from softening also showed narrow distributions with the PD in the range of 1.5 to 2.1.

As the extent of softening was increased, the average molecular weights of dextran60 dramatically moved to the greater elution volume, *i.e.*, to smaller molecular weights of organic matter. The change to smaller molecular weights also showed in the median values at each lime dose. The results clearly illustrate that softening preferentially removes the high molecular weight fraction of dextran. Since dextran showed similar results to NOM in softening and ultrafiltration behavior, we believe this same conclusion would hold for NOM.

SEM of Membrane Surfaces Fouled with the Raw Dextran

Scanning electron microscopy (SEM) was used to investigate directly the surfaces of fouled membranes. SEM images from the membrane fouled with raw Dextran60 are shown in Figure 5.11 at different resolutions from several places. SEM was performed at both low and high resolution. Images with a relatively low resolution are intended to illustrate the pattern of deposition on membrane surfaces, whereas those with a higher resolution help explain the interactions between foulant and membrane materials.

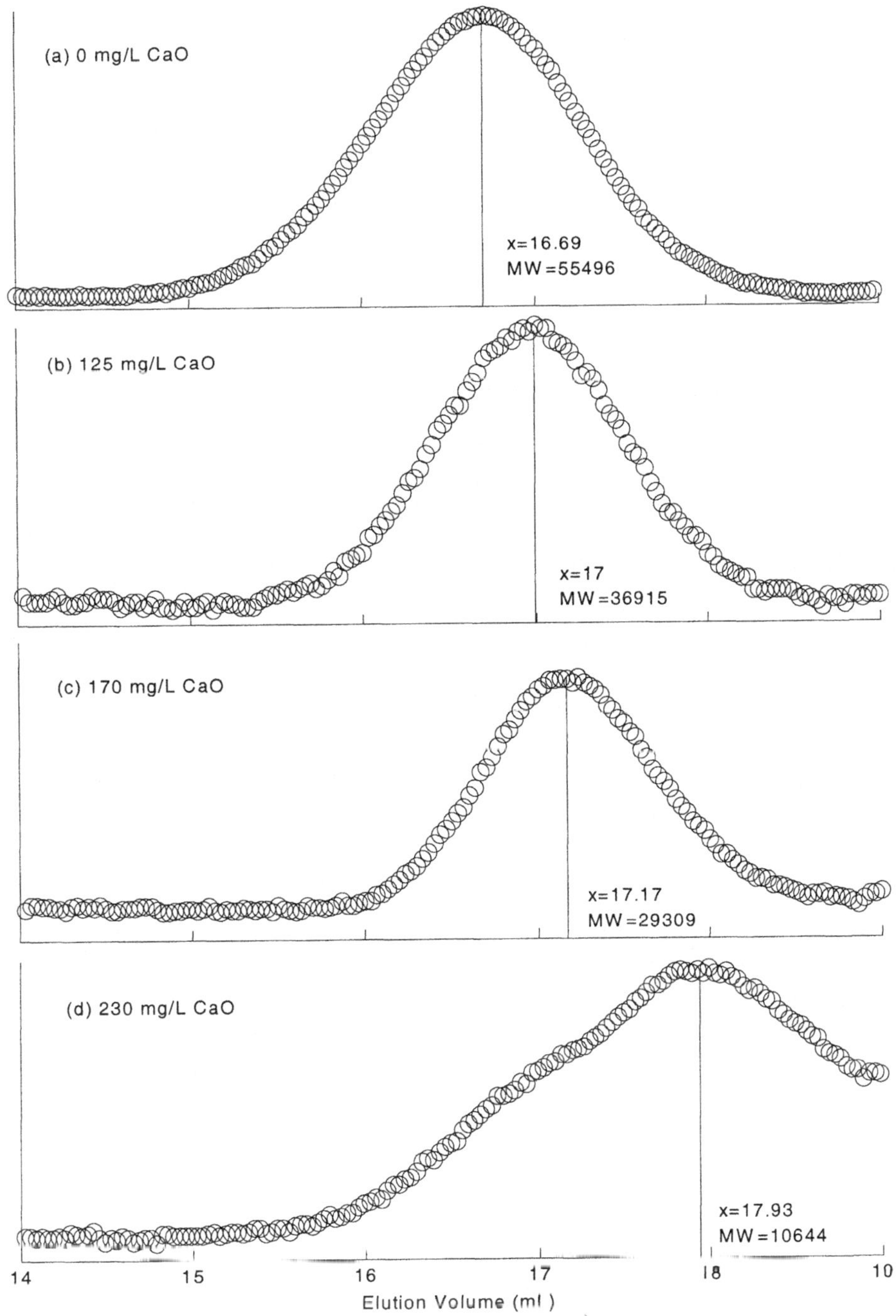

Figure 5.10 Molecular weight distribution of dextran after softening by SEC

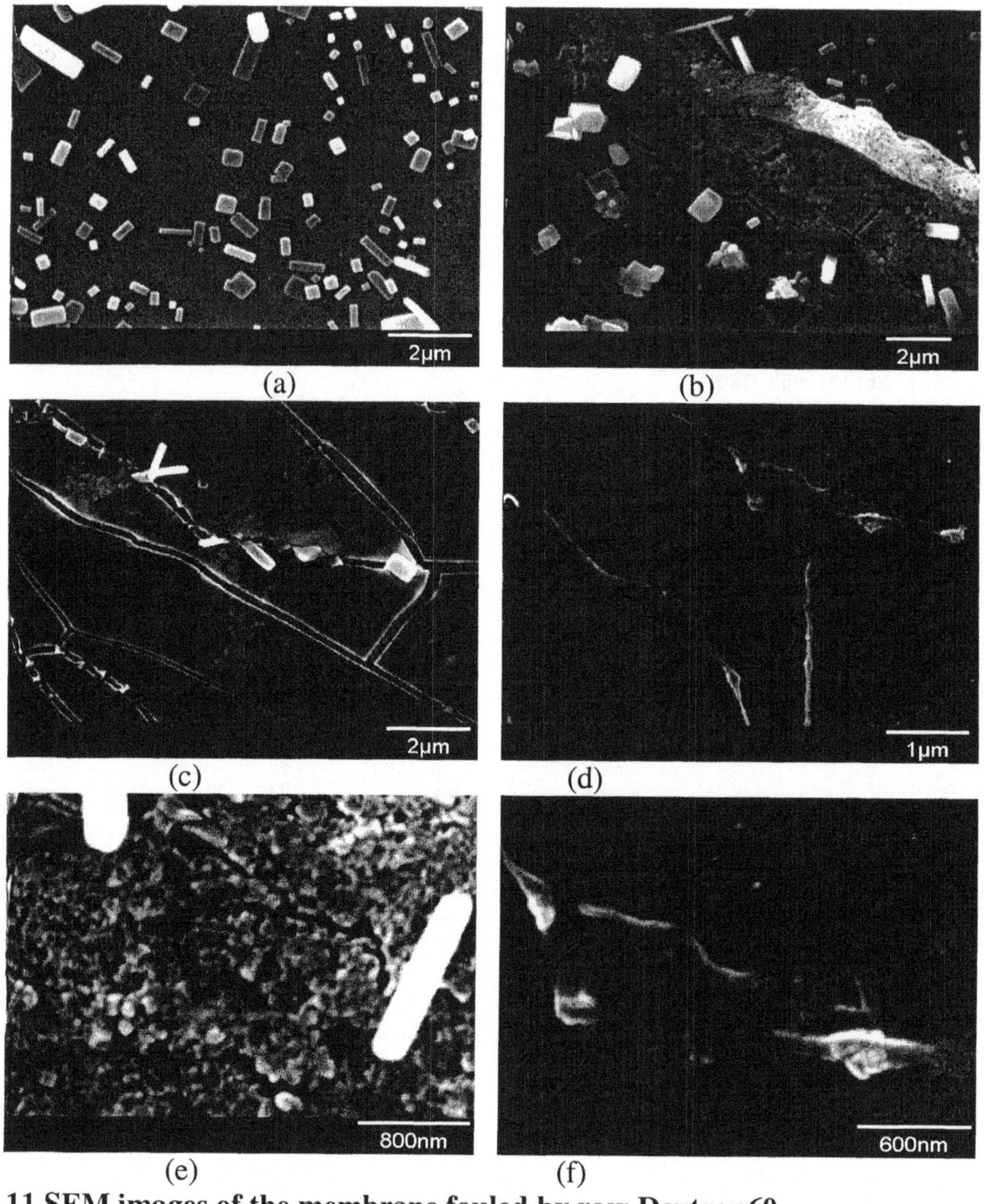

Figure 5.11 SEM images of the membrane fouled by raw Dextran60
([a], [b], and [c]: 10,000x; [d]:20,000x; [e]:35,000x; [f]:50,000x resolution)

The images from the low resolution (10k and 20k) of the membrane fouled by raw Dextran60 solution disclosed that several interesting events occurred on the membrane surfaces (shown in parts a, b, c, and d of Figure 5.11). Since the feed water was just raw Dextran60 solution in the hard water, no solid deposition was expected, but these figures make clear that substantial solid deposition actually occurred. Recall that the feed water is simulated Lake Austin water and is thus oversaturated with Ca^{2+} ions (with respect to $CaCO_3$). Precipitation can be initiated if a preferable condition such as a place with high surface energy like the presence of tiny particles or bumps in surfaces exists. The deposits on the membrane are apparently micrite, one of three main crystal morphologies of carbonate minerals. Micrite is formed from rapid precipitation and characterized by extremely fine, semi-opaque solids with the maximum crystal dimension less than 4 μm (Folk 1974). In comparison, most calcite has a dimension larger than 5 μm.

In addition, the four images from different positions at the low resolution reveal that deposition was non-homogeneous on the membrane surface. Image (a) shows a location with many deposits. Image (b) illustrates a location with a relatively small number of solids and a large single of deposit (indicating organic deposition). Image (c) shows a relatively clean location with many cracks, and image (d) magnifies the dextran attached to the membrane surface. These non-homogeneous deposits imply that some sites on membranes are more susceptible to either organic or inorganic fouling.

Image (c) dramatically showed the effects of cracks in the membrane. Due to the crack, more feed water filtered through this location and thus more foulants accumulated around and penetrated the crack. The TOC values from the run with this membrane showed that the TOC in the filtrate increased with operation time. The cause of the deterioration of the membrane is unclear now. It could be from either weakness of the membrane itself or some property of dextran, since this kind of deterioration was not detected from experiments with other water sources including Lake Austin water and alginic acid solution.

The images at the higher resolution were taken at 35k and 50k of magnification. The two images with the relatively high resolution (Figure 5.11 parts e and f) reveal that at least two different fouling mechanisms are associated with organic matter: a cake layer formation and a strong attachment (which could be a physical or chemical adsorption). Image (e) shows a local layer with thick deposition of organic matter (dextran) and image (f) shows a big lump of deposition as well as strong attachment at the edge of deposition. Although the shapes of deposition are precisely revealed through SEM, the properties of deposition must be investigated with other instruments.

SEM of Membrane Surfaces Fouled with Dextran After Softening

SEM images from the membrane fouled with the softened dextran solutions are shown in Figure 5.12 for the experiment at 125 mg/L lime dose. The images were taken at relatively low resolutions, i.e., 1k, 2k, and 10k since the deposits were relatively large. The images of foulants after softening treatment reveal that softening produced substantially different deposits than those from raw waters without softening.

At the lime dose of 125 mg/L, the images showed a lot of precipitates, which varied from needle-like shapes to well-defined rhombus shapes. The needle-like shapes seem to be elongated magnesian-calcite crystals, which are formed under a high Mg condition (Folk 1974). The formation of magnesian- calcite crystals stemmed from the standard softening condition, which maximizes Ca^{2+} removal but accomplishes little Mg^{2+} removal; therefore the ratio of soluble Mg^{2+}/Ca^{2+} is great. Surprisingly little organic deposition is seen in the SEM images considering the 23% DOC removal. Higher resolution or different locations on the membrane might have revealed more dextran, since non-homogenous fouling on the same membrane sheet was previously observed.

Images from the softened Dextran60 at 230 mg/L lime dose (Figure 5.13) show some isolated inorganic particles and their aggregates as well as inorganic particles associated within a fibril network of organic matter. These deposits were relatively widespread over the whole area of the membrane surface. The inorganic particles were mostly incorporated within organic matter. Therefore, sheet-like and less distinguished edges of depositions were revealed rather than well-defined rhombus shapes, the typical morphology of calcite; nonetheless, the rhombus shapes were detected within the organic matter.

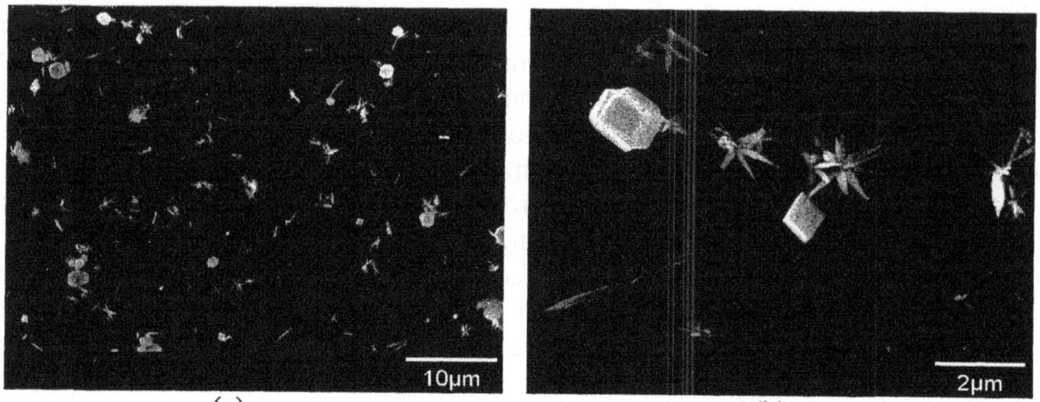

(a) (b)

Figure 5.12 SEM images of the membrane fouled by Dextran60 softened at 125 mg/L CaO ([a] 2,000x and [b] 10,000x resolution)

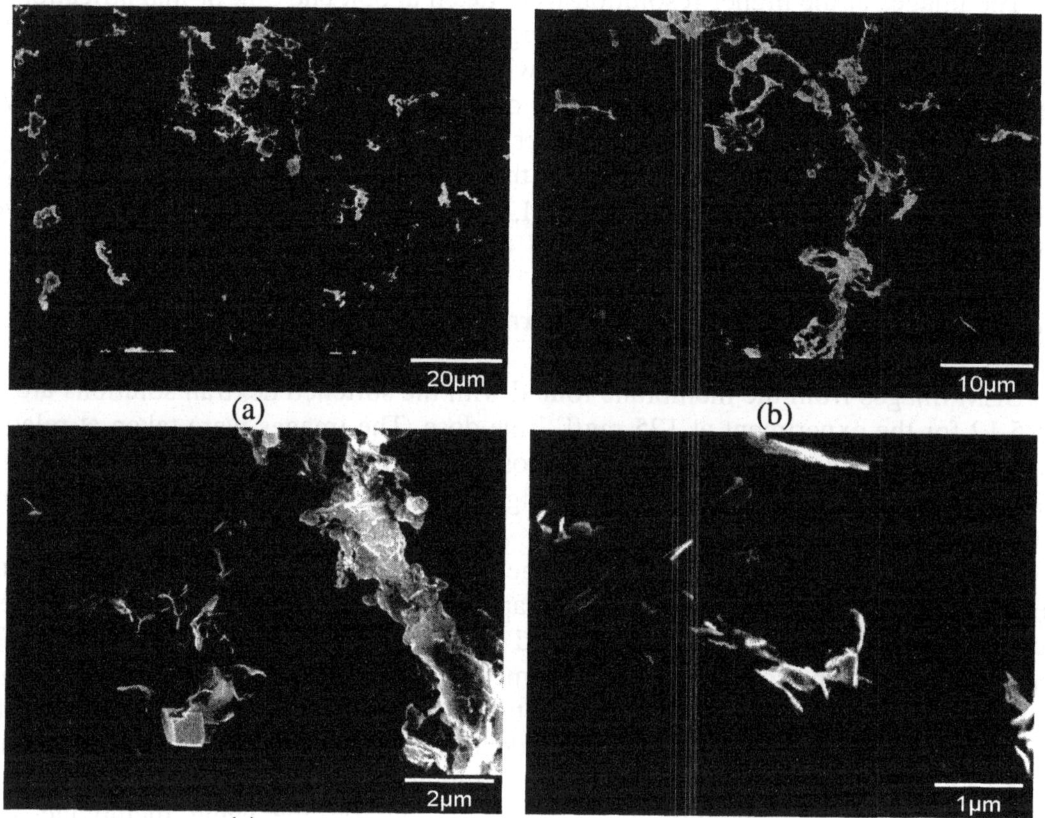

(c) (d)

Figure 5.13 SEM images of the membrane fouled by Dextran60 softened at 230 mg/L CaO (soda ash: 105 mg/L as CaO) ([a]: 1,000x; [b]: 2,000x; [c]: 10,000x; and [d]: 20,000x resolution)

XPS of Membrane Surfaces Fouled by Dextran

X-ray photoelectron spectroscopy was used to investigate the chemical composition and oxidation states of major elements in the topmost surfaces of fouled membranes. Survey spectra were obtained from a clean membrane and a membrane fouled with the softened Lake Austin water at the lime dose of 125 mg/L CaO. The high-resolution spectra were taken for the most substantial photoelectron peaks: C, O, S, Ca, and Mg.

The primary information obtained from XPS spectra is the binding energies from core level electrons and atomic concentrations of elements present on a surface. The binding energy is characteristic of the atomic core level from which the photoelectron was emitted (Munro and Singh 1993). The emitted electrons are referred to with a principle quantum number (i.e., 1, 2, 3......), orbital angular momentum (i.e., s, p, d, f.....), and total angular momentum (i.e., ...). To determine line energies accurately, the voltage scale of the instrument must be precisely calculated. During the analysis, samples tend to acquire a steady-state charge. One of methods to calibrate charging effects is to measure the position of the C(1s) line from hydrocarbons that are nearly always present in the samples (Moulder et al. 1992). In this research, the carbon in the membrane material, i.e., polysulfone, was used as a calibration line. Therefore, any shift from this value could be taken as a measure of the static charge, and the sample charging in this research was approximately 2.4 eV.

The composition of the main elements in the membrane surfaces fouled with dextran solutions is presented in Table5.9. The elemental composition from the membrane fouled with the softened water at lime dose of 125 mg/L CaO was quite similar to that fouled with the raw dextran. However, the membrane fouled at 230 mg/L CaO showed a lower C(1s) and a greater O(1s) composition than the membrane fouled with the raw dextran. As lime doses were increased, the deposited Ca^{2+} and Mg^{2+} clearly increased. For instance, Mg^{2+} was not detected at the raw dextran solution because it passed through the membrane. Then, the Mg^{2+} composition increased to 3.7% at the high lime dose, i.e., 230 mg/L as CaO reflecting the precipitation of $Mg(OH)_2$. Therefore, the increases in the inorganic compounds reflect the precipitation of $CaCO_3$ and $Mg(OH)_2$ on the membrane surface, even though most of precipitates were removed by settling prior to application with membrane. These spectroscopy results are consistent with the SEM images, which showed substantial amounts of solid deposition on the fouled membrane after softening.

In comparison, the values of clean and fouled membranes from other research are shown in Table 5.10. Since XPS is a relatively novel instrument that has not often been applied to membrane processes in drinking water systems, there is little and scattered information about the atomic composition of the specific membrane materials. More research in XPS analysis should be performed to accumulate knowledge of the atomic composition of membranes and deposits on membranes.

Table 5.9
Atomic composition: effects of softening Dextran60 at different lime doses

Lime dose (mg/L CaO)	C(1s) (%)	O(1s) (%)	S(2p) (%)	Ca(2p) (%)	Mg(2s) (%)
0	72.5	21.3	5.9	0.4	NA
125	72.5	20.9	5.6	0.9	0.1
230	66.2	25.3	4.8	1.0	3.7

1s: s orbit in the 1st energy level, 2p: p orbit in the 2nd energy level; 2s: s orbit in the 2nd energy level

Table 5.10
Atomic composition of membranes in the literatures

Membrane	C(1s) (%)	O(1s) (%)	S(2p) (%)	Reference
Polysulfone	81.3	15.3	3.4	Oldani and Schock, 1989
Poly (*p*-phenylene ether sulfone)	74.3	18.8	6.8	
Polysulfone*	70.4	27.0	1.7	Jucker and Clark, 1994
Urea formaldehyde polysulfone**	82.0	14.0	3.0	Palacio *et al.* 2001
Polysulfone	72.3	21.7	6.0	This research, 2002

* Measured after 8.3 mg/L of humic substance adsorbed and the Ca accounts for 0.97% of the atomic composition.
** Nitrogen (N) was less than 2% of the membrane composition

Further, the spectrum indicates the state of valence electrons, which are usually involved in chemical bonding. Chemical bonding of valence electrons influences the core-level binding energies in an atom, and thus produces a range of binding energies (chemical shift). Unfortunately, the range of chemical shift for an element is usually smaller than the energy resolution of the XPS instrument; therefore several peaks could be superimposed in the range. A Gaussian line shape for a component is usually used for curve fitting of peaks (Briggs and Riviere 1983, Munro and Singh 1993). Figure 5.14 shows the spectrum of C(1s) from the clean membrane in this research, along with an example of curve fitting. Two binding energies are identified from the curve fitting: 285.1 eV and 286.8 eV. The binding energy of 285.1 eV is the characteristic of atomic carbon and the binding energy of 286.8 eV is from the structure of the polysulfone membrane. For this research, the Excel solver routine was used to find the set of gaussian distributions that best fit the raw data, i.e., minimized the residual sum of squares of the differences between the model and experimental data. The figure makes it clear that a shoulder in the experimental results stems from the superposition of two Gaussian curves. Patel et al. (1988) found that polysulfone exhibited three peaks corresponding to its various chemical bonds, i.e., a peak of 284.8 eV is for carbon atoms not attached to O or S, a peak of 286.1 eV is for carbon attached to S, and a peak of 286.6 eV is for carbon attached to O.

The spectra of C(1s), O(1s), and S(2p) from the clean membrane in this research were analyzed by fitting Gaussian curves to the data. From the results, the width and position of the peak of each element was fixed and used as a standard for the other membranes, so only the intensity of the peak was changed to fit a graph from fouled membranes. Sometimes, two peaks were not adequate to describe the spectra, so that a third peak had to be considered. This third peak was required for membranes fouled with alginic acid, Lake Austin water, and Missouri River water.

As shown in the results of the atomic composition (Table 5.11), the Ca^{2+} and Mg^{2+} became detectable in the fouling layer with increasing lime doses. The characteristic binding energies of Ca are 347.2 ± 0.1 eV for Ca($2p^{3/2}$) and 351.0 ± 0.1 eV for Ca($2p^{1/2}$). Jucker and Clark (1994) also found that the binding energy of Ca($2p^{3/2}$) was in the range of 349.2 – 349.7 eV when humic acid was adsorbed on a polysulfone membrane. The authors assumed that calcium was bound in some manner to the humic acid molecules, which led to the change of the binding energy. In the experiment with the lime dose of 125 mg/L CaO, two peaks were identified: 349.0 for Ca($2p^{3/2}$) and 352.3 eV for Ca($2p^{1/2}$). These results suggest that Ca was possibly incorporated into the dextran to foul the membrane at that lime dose. At the lime dose of 230 mg/L CaO, only one peak of Ca(2p) was detected at approximately 352.8 eV, which might indicate more rigorous interactions of Ca with other elements such as dextran to yield the higher binding energy.

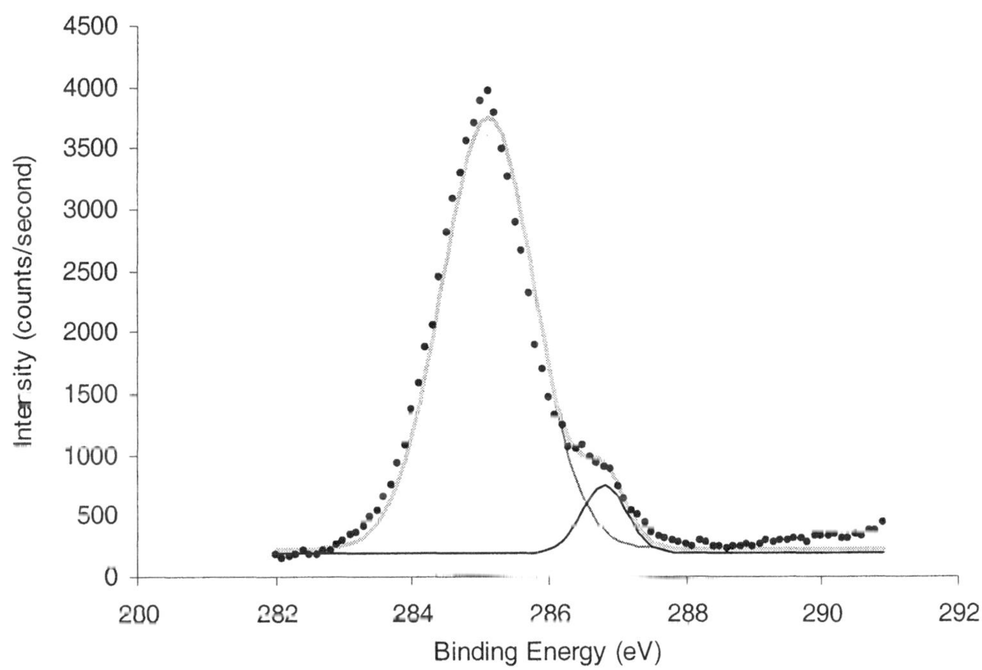

Figure 5.14 C(1s) spectrum from the clean membrane (polysulfone)

Table 5.11
Binding energy of each atomic composition of the membranes fouled by Dextran60

Atom	Survey range	Typical value*	Lime dose (mg/L)		
			0	125	230
O(1s)	525-545	531	531.9	531.9	532.2
S(2p)	158-178	164	167.9	168.1	168.6
Ca(2p)	342-362	347, 351	NA	349, 352.3	352.8
Mg(2s)	84-104	89	NA	NA	90.6

* adapted from Moulder J. F. et al., (1992), Handbook of X-ray Photoelectron spectroscopy.
Note: The binding energy of C(1s) fixed at 285.1 eV.

Detailed spectra of C, O, Ca, and Mg in the experiments with different lime doses are shown in Figure 5.15. The spectra of S(2p) were not shown because of small changes in the spectra regardless of the lime dose. As discussed earlier, the main peak of C(1s) was fixed at 285.1 eV to compensate for charging effects from different samples. The spectra of C(1s) (Figure 5.15(a)) showed that the intensity of the binding energy at 286.7 – 286.8 eV increased after ultrafiltration of the raw and the softened dextran solutions compared to the clean membrane. Considering that carbon atoms attached to oxygen exhibit the binding energy at approximately 286.8 eV, the increased intensity at this binding energy was an indication of organic fouling by dextran. However, the intensity was approximately the same for the two different degrees of softening.

Figure 5.15 (b) shows that the spectra of O(1s) were changed little except in the experiment with a lime dose of 230 mg/L CaO. The binding energy of O(1s) was reported to display at 531.7 for atomic oxygen, 532.3 eV for O – S bond, and 533.5 – 533.7 eV for O – C bond (Moulder *et al.* 1992, NIST 2002). The fact that little change of intensity occurred at the binding energy of O(1s) at 533.5 eV in the experiments with lime doses of 0 and 125 mg/L CaO was consistent with the results from the C(1s) spectra at the binding energy of 286.8 eV. However, the intensity of O(1s) at 533.5 eV increased in the experiment with lime dose of 230 mg/L CaO, which could not stem from the C – O interaction since the C(1s) spectra was not changed at the lime dose. The spectra of Ca(2p) and Mg(2s) can give an explanation of the increased intensity at the lime dose of 230 mg/L CaO. As shown in Figure 5.16(c) and (d), the intensities of Ca and Mg in the experiment with lime dose of 230 mg/L CaO were significantly increased compared to other experiments. Therefore, the increased intensity at 230 mg/L CaO was likely from the precipitation of $CaCO_3$ or $Mg(OH)_2$. However, the characteristic binding energies of O(1s) with $CaCO_3$ (531.2 – 531.4 eV) and MgO (532.1 eV) are far from the binding energy with increased intensity at 230 mg/L CaO, *i.e.*, 533.5 eV (NIST 2002). This result might imply that the increased intensity at the experiment with the lime dose of 230 mg/L CaO is not directly from $CaCO_3$ or $Mg(OH)_2$ deposits, but from some reactions between inorganics (Ca and Mg) with oxygen in the membrane. This idea was supported from the spectrum of Ca(2p), which changed substantially in its characteristic binding energy in the experiment with the highest lime dose. The binding energy of Ca(2p) is reported at 347.2 eV for Ca($2p^{2/3}$) and 351.1 eV for

$Ca(2p^{1/2})$, which is close to the results from the experiment with lime dose of 125 mg/L CaO. However, the location of a peak in the experiment with lime dose of 230 mg/L appeared at approximately 353 eV, which was quite different from the reported values. This might indicate that more than one Ca chemical state was present.

5.2.3 Investigation of Alginic Acid on Membrane Fouling

Effects of Extent of Softening on Fouling by Alginic Acid

Another polysaccharide compound, alginic acid, was investigated for effects of the degree of softening pretreatment on membrane fouling. The alginic acid in this research has a nominal molecular weight of 12~80 kDa according to the manufacturer (Aldrich, St. Louis, MO). The softening process was the same as in the experiments with dextran.

Table 5.12 summarizes operational conditions in each test with 4 mg/L C of alginic acid at three different lime doses. The operational conditions including crossflow velocities and TMPs were maintained in the designated values during the experiments. In most cases, the clean water specific fluxes were quite consistent between 1.6 and 1.8 L/m^2-hr-kPa.

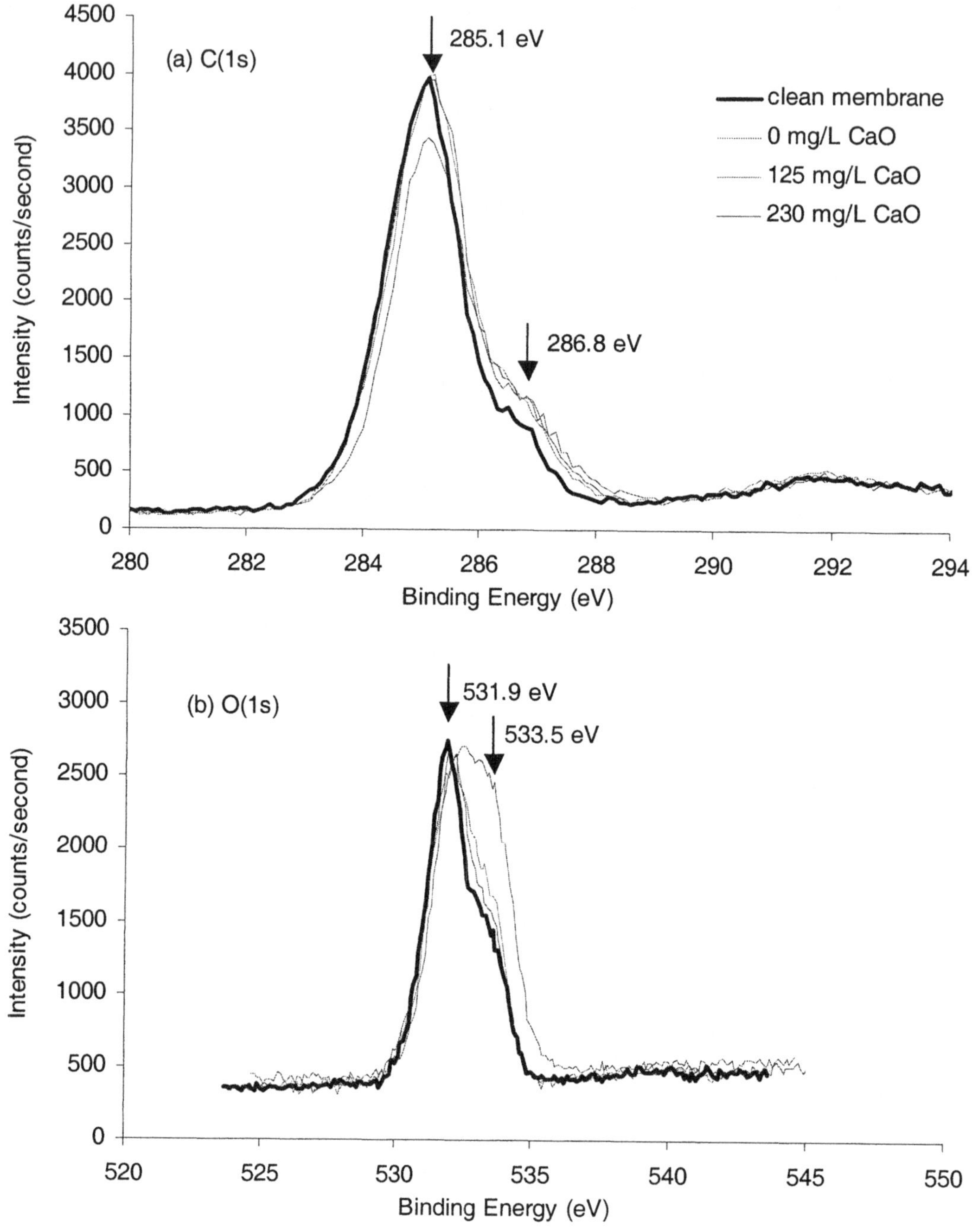

Figure 5.15 Spectra of the membrane fouled by Dextran60 at different lime doses

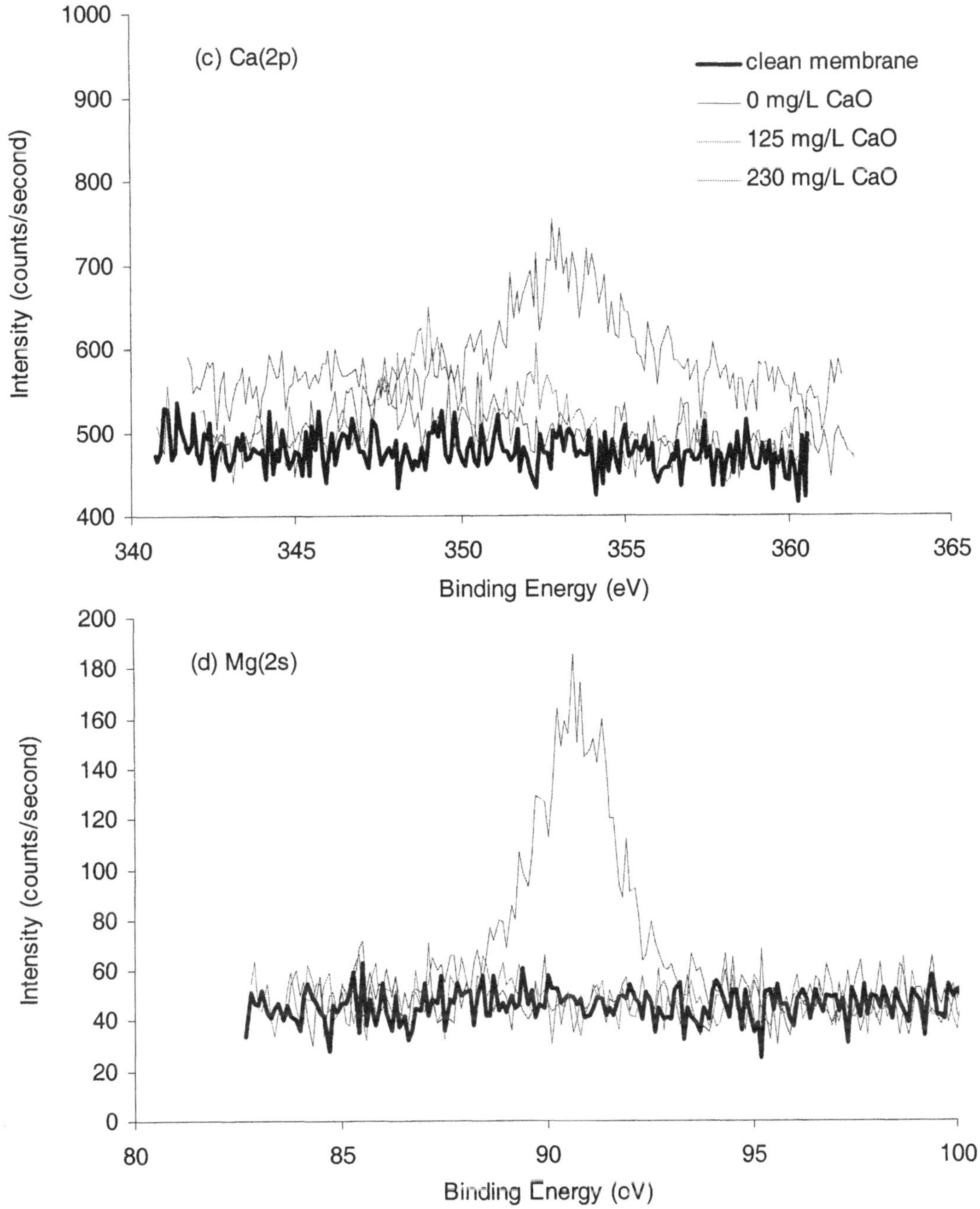

Figure 5.15 Continued

Table 5.12
Operational conditions for the experiments with 4 mg/L of alginic acid at different lime doses

Lime (mg/L CaO)	Soda ash (mg/L CaO)	TMP (kPa)	Crossflow Velocity (cm/s)	Clean water specific flux, J_{so} (L/m^2-hr-kPa)
0	-	91.0 - 95.2	10.3	1.79
125	-	91.0 - 99.3	10.5	1.55
170	45	84.8 - 93.8	9.92	1.76
230	105	85.5 - 91.0	10.2	1.73

Table 5.13
Water quality of experiments with alginic acid (DOC: 4 mg/L C)

Lime (mg/L CaO)	Soda ash (mg/L CaO)		pH	Turbidity (NTU)	Ca^{2+} (mg/L)	Mg^{2+} (mg/L)	DOC (mg/L)
0	0	Feed	8.41	NA	70.4	16.7	4.7
		Filtrate	8.48		66.1	16.9	0.3
125	0	Raw	8.28	NA	54.8	14.8	4.2
		Softened	10.73	27.3	20.5	13.5	3.9
		Feed	10.54	3.78	19.3	14.1	3.8
		Filtrate	10.53	0.020	15.6	13.7	D.L.
170	45	Raw	8.42	NA	50.9	14.8	3.9
		Softened	11.06	14.5	16.5	5.1	NA
		Feed	10.88	4.25	12.6	3.9	2.9
		Filtrate	10.80	0.016	9.4	2.3	0.3
230	105	Raw	8.34	NA	50.0	16.2	3.4
		Softened	11.48	31.2	12.9	3.0	1.9
		Feed	11.36	3.57	5.5	2.0	1.5
		Filtrate	11.37	0.015	2.7	0.3	0.6

D.L.: lower than detection limit

Table 5.13 presents water quality at various stages of all of the experiments using the alginic acid solution as a raw water. In general, turbidity and inorganic constituents such as pH, Ca^{2+}, and Mg^{2+} were reasonably consistent with the previous screening test and were similar to the Lake Austin water. As the only exception to the consistency, the calcium concentration in the experiment with raw alginic acid was slightly higher than that in Lake Austin water.

As expected, the DOC removal substantially increased (from 7% to 44%) as the lime dose was increased from 125 mg/L to 230 mg/L CaO. However, the removal efficiency was smaller than that from the screening test, where it increased from 52% to 89% with the same change of lime doses. The reason stems from differences in experimental procedures. During the screening test, all softened waters were filtered through 0.45 μm membrane filters. However, the softened waters during these experiments were usually settled overnight and then the supernatant siphoned through a Tygon tube to collect the feed water for the membrane experiment. To test whether the solid/liquid separation methods made the difference, a simple experiment was performed. Raw alginic acid solutions (approximately 10 mg/L C) were prepared with distilled/deionized water and with the synthetic inorganic water (i.e., a hard water containing the same amounts of Ca^{2+} and Mg^{2+} as Lake Austin water). The organic content was then measured before and after 0.45 μm-membrane filtration. The results indicated that the alginic acid was significantly removed (approximately 57% of DOC) in the filtration process when it was prepared in the synthetic inorganic water. Little removal occurred in the solution made with distilled/deionized water. The presence of hardness ions might induce complexation or precipitation of alginic acid under the high ionic strength condition. Therefore, the greater removal in the screening test was an overestimate of the removal efficiency of alginic acid in softening.

The molecular weight of the raw alginic acid in the synthetic inorganic water was analyzed with size exclusion chromatography as shown in Figure 5.16. The result shows a wide distribution of molecular weight with a median value of 519 kDa, considerably higher than the nominal molecular weight reported by the manufacturer (12-80 kDa). The degree of polydispersity of alginic acid (i.e., 6.2) also confirmed the wide distribution. The SEC analysis confirmed that the alginic acid formed larger molecules by complexation with hardness ions.

The removal of alginic acid (more than 90%) as well as turbidity (almost 100% removal) was superb in these ultrafiltration experiments. This removal was predictable because the nominal molecular weight of the alginic acid by the manufacturer was in the range of 12-80 kDa, and the molecular weight distribution measured by SEC was much bigger than the nominal membrane pore size. The high removal of alginic acid by ultrafiltration could be a great drawback to use it as a NOM surrogate because, in general, NOM is marginally removed by ultrafiltration.

It is also notable that some Ca^{2+} and Mg^{2+} in the feed water was removed by ultrafiltration when softening was applied, which was not shown with the dextran solution. It is likely that the Ca^{2+} and Mg^{2+}, which were associated with alginic acid by complexation or precipitation, were removed from the solution by ultrafiltration.

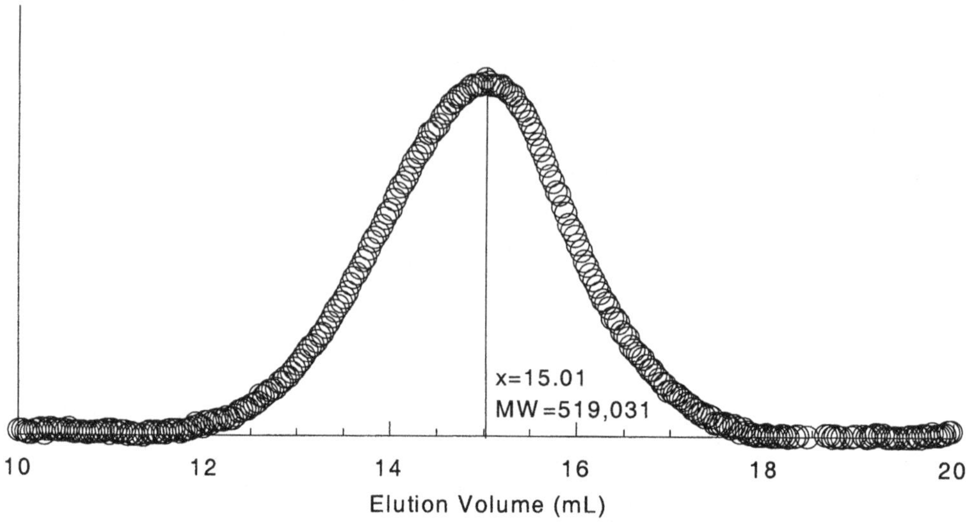

Figure 5.16 Molecular weight distribution of the raw alginic acid

The flux decline with the alginic acid at different lime doses is shown in Figure 5.17(a). The flux improved as lime doses were increased, which indicates that softening effectively removes a portion of alginic acid that causes fouling in ultrafiltration. Recall (from Table 5.13) that the DOC in the feed water decreased with an increasing extent of softening, so that the improved flux appears directly related to the decreased DOC concentration. Recall that the dextran solution showed little flux improvement between 170 and 230 mg/L doses of lime, but that result might have occurred because very little flux decline occurred after softening at the 170 mg/L of lime.

To test quantitatively whether the flux decline is directly caused by the overall DOC in these alginic acid solutions, the results are re-plotted in Figure 5.17(b), based on the cumulative mass of DOC fed to the membrane (per unit area), rather than the cumulative volume of water. This measure was obtained by multiplying the cumulative water production by the feed DOC concentration in each case. The graph is quite interesting. The graphs after softening nearly (but not quite) collapse into one line, but that line is quite different from the no softening results. The difference between the no softening and the softening results could come from three things--pH differences leading to different behavior of the organics interacting with the membrane surface, more particles being present in the softening cases (from the precipitation of inorganic solids), or a preferential (first) removal in softening of that portion of alginic acid that causes the most fouling (or a combination of all of these). Other results from this research suggest that the first two explanations are unlikely. The fact that the three remaining curves show some improvement of flux with increasing softening is consistent with the idea that the portion of alginic acid that causes the most flux decline is removed first in softening. Although not shown, the dextran results were also plotted in the same way, and these graphs did not collapse together at all, again meaning that the portions of dextran that caused the most fouling were removed preferentially in softening.

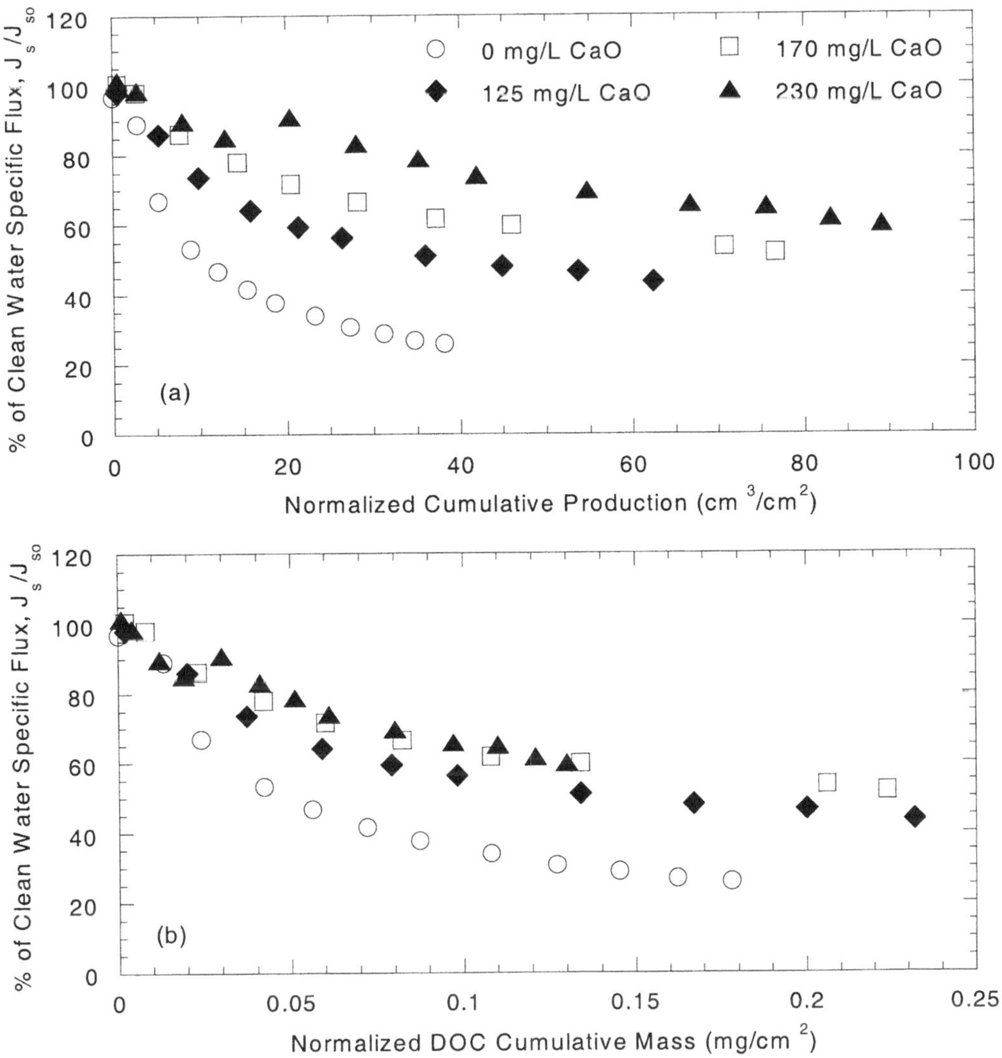

Figure 5.17 Flux decline: effects of softening alginic acid (4mg/L C) at different lime doses (a) normalized with cumulative water production, and (b) normalized with cumulative DOC mass

Efficiency of Each Cleaning Method on Alginic Acid

Results of the efficiency of the various cleaning methods for membranes fouled with the alginic acid are shown in Figure 5.18. In general, surface wash was very effective in recovering the lost flux in all experiments; the best result was a recovery of approximately 70% of the lost flux for the membrane fouled with the softened water at 230 mg/L of lime. In each case, the caustic wash and the subsequent acid wash improved the recovery a few more percent. Considering that the caustic wash is intended to remove organic foulants, the effective cleaning by the acidic wash was relatively significant. It is consistent with the idea that inorganics complex with alginic acid and therefore play a role in membrane fouling. In general, all three cleaning methods applied on the fouled membranes were effective.

SEM of Membrane Surfaces Fouled with Alginic Acid

Figure 5.19 shows SEM images from the membrane fouled with the raw alginic acid at three different resolutions. All images show a lot of fibrils, which are a typical indication of high polysaccharide content. The shapes are extended coils rather than a rigid rod, which makes sense since alginic acid has a high charge density so that the functional groups in alginic acid repel each other. In addition, it should be noted that little solid deposition is shown compared to the images from the membrane fouled with the raw Dextran60.

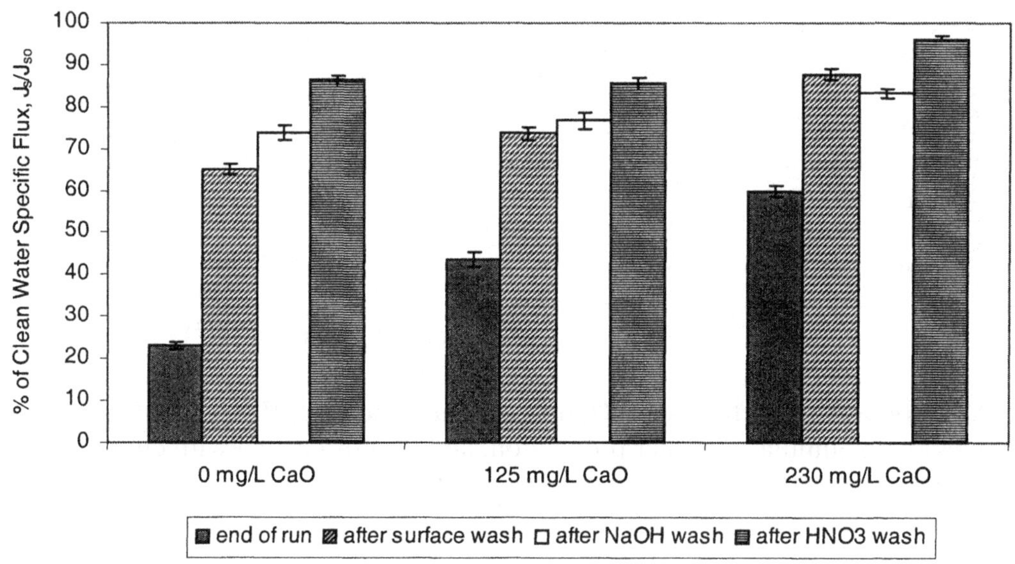

Figure 5.18 Membrane cleaning: alginic acid solution at different lime doses (4 mg/L C)

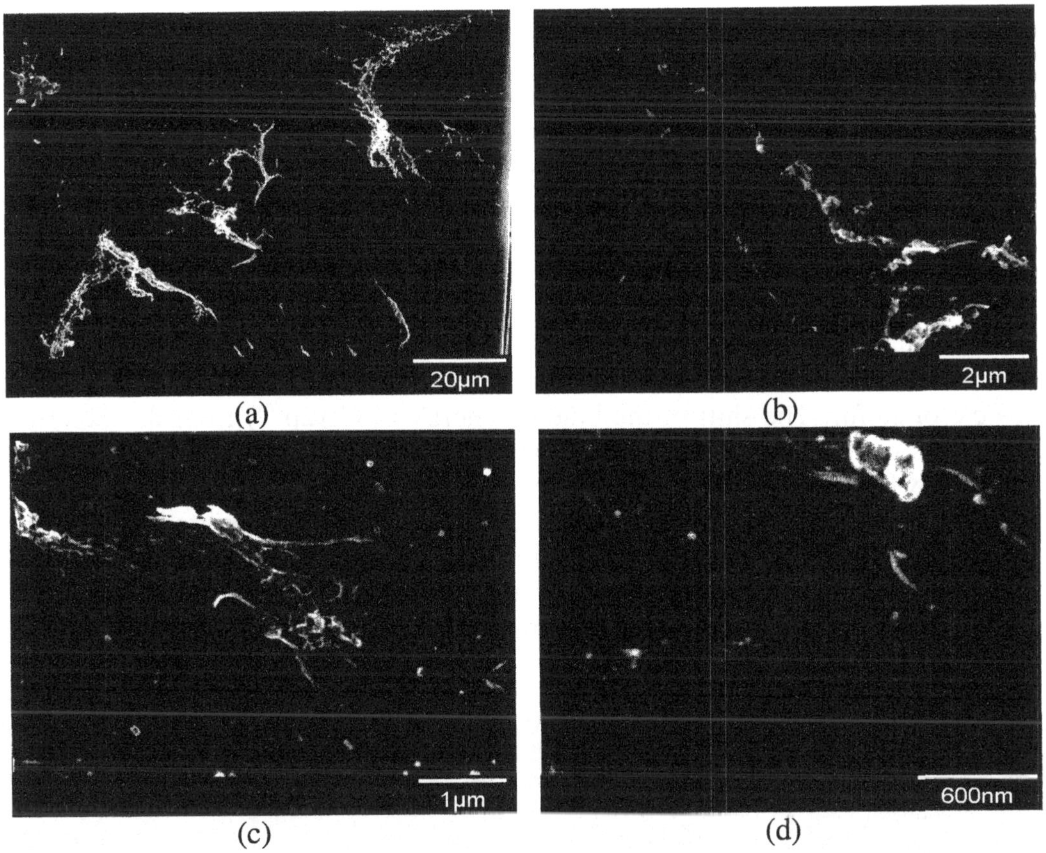

Figure 5.19 SEM images of the membrane fouled by raw alginic acid
([a]: 1,000x; [b]: 10,000x; [c]: 20,000x; [f]: 50,000x resolution)

The SEM images from the membrane fouled with the softened alginic acid at 125 mg/L CaO are shown in Figure 5.20. As expected, the images after softening show more inorganic particles, which are isolated or embedded within thread-like shapes of organic matter. The inorganic solids, which are presumably $CaCO_3$, show spherical shapes of precipitates, which are usually formed under a relatively high concentration of Mg^{2+} (Folk 1974). At the lime dose of 125 mg/L, the calcium removal is optimized but the magnesium removal is very limited since little $Mg(OH)_2$ precipitation occurs in the range of pH. In addition, there is some suggestion in the results shown in Figure 5.20 (b) that organic matter acts like a glue so inorganic particles are sitting on the top of the organic deposits. These results are consistent with those by Mallevialle, Anselme, and Marsigny (1989).

XPS of Membrane Surfaces Fouled by Alginic Acid

As shown in Table 5.14, the composition of the main elements in the membrane surfaces fouled with alginic acid solutions was also determined. The composition of O(1s) became a larger portion on the membrane fouled with the raw alginic acid, *i.e.*, 27%, compared to those with the raw dextran, *i.e.*, 21%. The increased portion of oxygen stems from the fact that alginic acid has a molecular structure (a carboxylic acid functional group (-COOH)) with more oxygen than dextran. As lime doses increased, the composition of the species after softening the alginic acid solution became close to those with dextran.

The Ca and Mg compositions in the membrane fouled with the alginic acid were relatively constant, from 1.1 to 1.3% for Ca and from 0.2 to 0.7 % for Mg, regardless of the degree of softening. Recall that the membranes fouled with the dextran solutions showed that the inorganic composition increased with the degree of softening. The constant inorganic composition here is again consistent with the idea that the raw alginic acid is complexed with inorganic matter to some extent regardless of the degree of softening.

The binding energy of the main peak of each component from the membranes fouled with alginic acid is summarized in Table 5.15, along with its typical value. The binding energy of O and S were close to the characteristics of the elements in polysulfone, i.e., 531.7 eV of oxygen and 167.9 eV of sulfur. The shift of the binding energy of Ca($2p^{2/3}$) from the experiments with alginic acid was not as great as those with dextran. The binding energies of Ca($2p^{2/3}$) was 347.7 ± 0.1 eV, slightly higher than the characteristic binding energy of Ca($2p^{2/3}$) (*i.e.*, 347.2 ± 0.1 eV). The Ca($2p^{2/3}$) in the calcium hydroxide with polyacrylic acid was reported to have the binding energy at 347.5 eV (NIST 2002). Therefore, the shift of the binding energy of Ca($2p^{2/3}$) likely reflects the chemical changes around Ca as the element complexed with alginic acid. The binding energy of Mg(2s) could not be determined because the signal was weak compared to background noise.

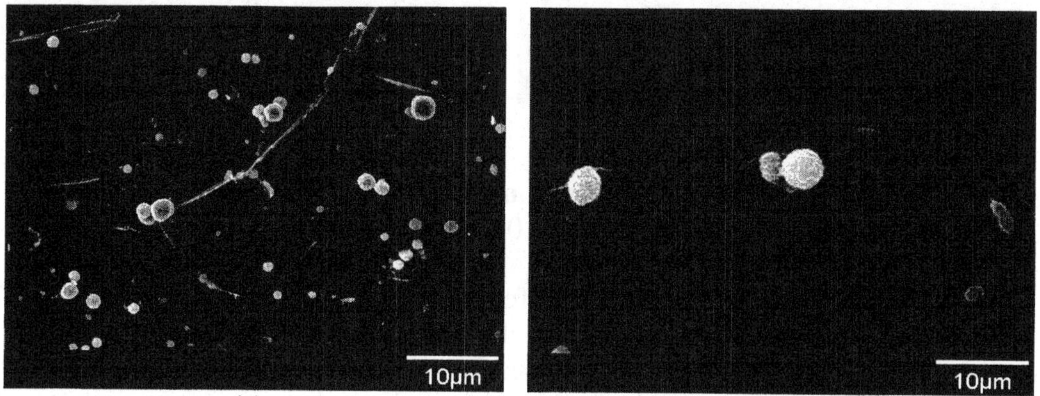

(a) (b)

Figure 5.20 SEM images of the membrane fouled by the alginic acid softened at 125 mg/L CaO (2,000x resolution)

Table 5.14
Atomic composition: effects of softening alginic acid at different lime doses

Lime dose (mg/L CaO)	C(1s) (%)	O(1s) (%)	S(2p) (%)	Ca(2p) (%)	Mg(2s) (%)
0	68.2	26.5	3.5	1.3	0.5
125	69.5	24.5	4.3	1.1	0.7
170	71.4	21.7	5.4	1.3	0.2

Table 5.15
Binding energies of each component of the membranes fouled by alginic acid

Atom	Survey range	Typical value*	Lime dose (mg/L)		
			0	125	170
O(1s)	525-545	531	531.9	531.7	531.9
S(2p)	158-178	164	168.5	168.2	168.3
Ca(2p)	342-362	347, 351	347.6, 352.1	347.8, 350.6	347.7, 351.5
Mg(2s)	84-104	89	NA	90.0, 101.6	101.6

* adapted from Moulder J. F. et al. (1992), Handbook of X-ray Photoelectron spectroscopy.
Note: The binding energy of C(1s) fixed at 285.1 eV.

Detailed spectra of C(1s) and O(1s) from the experiments with alginic acid at different lime doses are shown in Figure 5.21. The spectra of Ca(2p) and Mg(2s) are not shown because the changes in the spectra were quite small regardless of lime doses. As shown Figure 5.21 (a), C(1s) spectra from the membrane fouled with the raw alginic acid and that from the softening at the lime dose 125 mg/L CaO showed much broader shoulders than those from the clean membrane and the membrane fouled with dextran (in Figure 5.14 and Figure 5.15). Due to the tailed curve shapes of the spectra, two peaks were not sufficient to describe the spectra shapes; therefore, three peaks were identified with binding energies of 285.1, 286.5, and 288.2 eV.

Briggs (1983) suggested that oxygen induces shifts to higher binding energy by ~1.5 eV per C – O bond for C(1s) binding energy. The author illustrated an example from the spectra of low-density polyethylene (LDPE). The main peak of the C(1s) atom was at 285 eV. The other three peaks in the shoulder were ascribed to C – O (e.g. alcohol, ether, ester) at ~286.5 eV, > C = O (e.g. aldehyde, ketone) at 288.0 eV, and – COO (carboxylic acid or ester) at 289.5 eV. This example clearly showed that the binding energy was shifted by 1.5 eV as C – O bonds increased. The results from the membrane fouled with alginic acid also showed that the binding energy shifted by approximately a 1.5 eV interval from 285.1 eV. The third peak at 288.2 eV, therefore, indicated the carboxylic functional group in the alginic acid. The peak at the 288.2 eV was not detected for the experiment from the softened alginic acid at the lime dose of 170 mg/L.

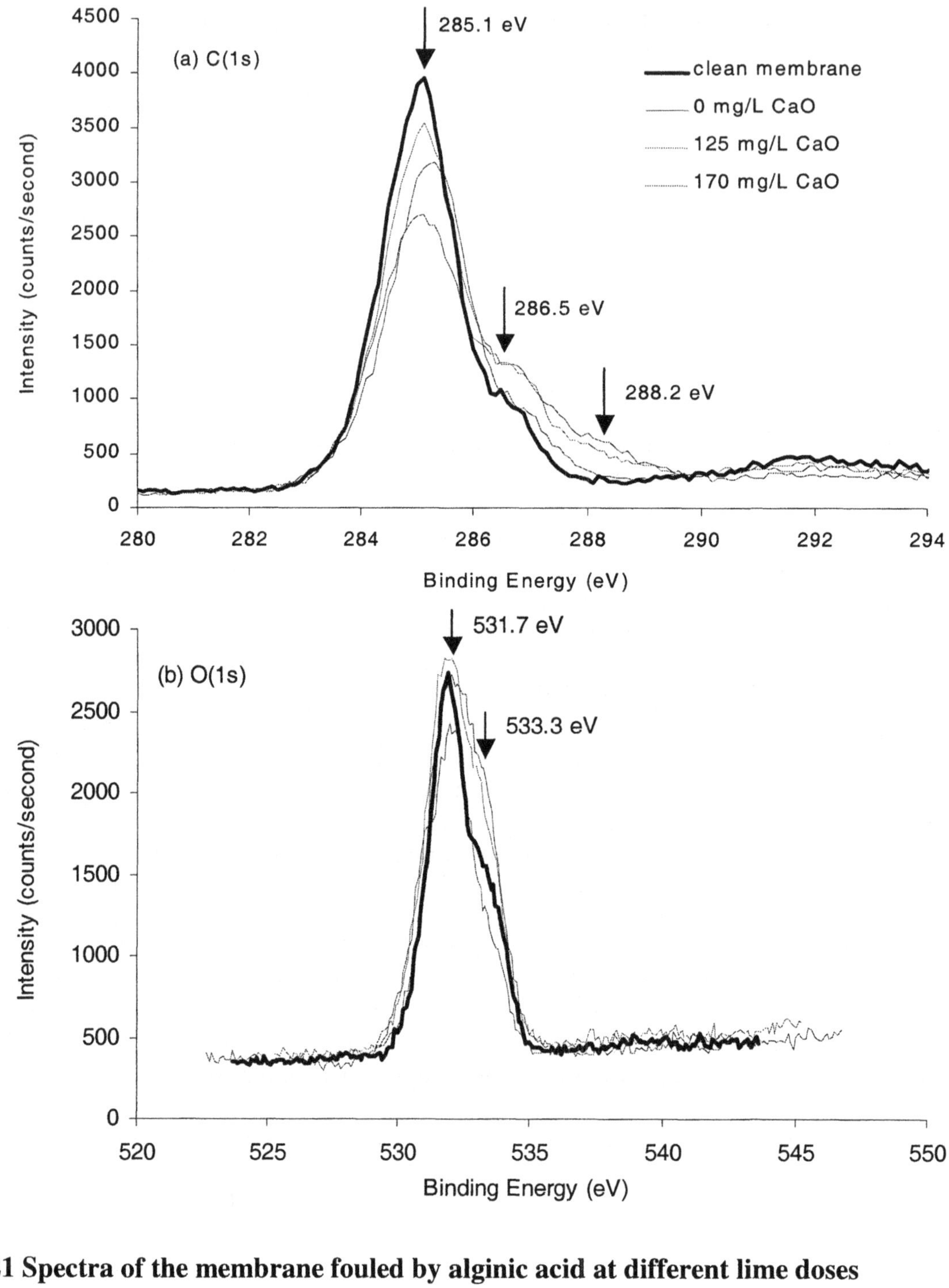

Figure 5.21 Spectra of the membrane fouled by alginic acid at different lime doses

In addition, the increased intensity of C(1s) at the binding energies of 286.5 and 288.2 eV was consistent with the results from the spectra of O(1s) as shown in Figure 5.21(b). The increased intensity at the binding energy of 533.5 eV in the experiments with the two lime doses (0 and 125 mg/L CaO) corresponded to the increased C – O bond in the environment of O(1s). It was also reflected in the greater composition of O(1s) at the lime doses (shown in Table 5.14).

5.3 PARTICLE FOULING

Particulate matter, which can be substantially removed in the softening process, can be a major foulant in membrane processes. Therefore, effects of particles on membrane fouling were investigated in this research with two turbidity concentrations, approximately 50 and 500 NTU, using a synthetic inorganic water with only clay particles to eliminate the effects of NOM.

5.3.1 Mass Transport of Particles in Ultrafiltration

When particles enter a membrane system, numerous transport mechanisms affect the velocity of each particle at each location. The transport mechanisms that carry particles toward the membrane surface include convection by permeation drag and settling by gravity. The transport mechanisms away from the membrane (i.e., back transport) in the crossflow mode are Brownian diffusion, shear-induced diffusion, and inertial lift (Belfort, Davis, and Zydney 1994; Wiesner and Aptel 1996). Under a typical operational condition in ultrafiltration, a particular range of particles is most susceptible to deposit on the membrane surface or block the membrane pore. Figure 5.22 presents the transport velocity by each back transport mechanism under the operational conditions in this research. The velocity by the Brownian diffusion is inversely proportional to particle diameter, and the velocities by both the shear-induced diffusion and the inertial lift are directly proportional to the particle diameter. The results in Figure 5.22 show that particles in the range between 0.3 and 1 μm ($-0.5 < \log d_p < 0$) had the minimal back transport velocity in the membrane system. Therefore, particles with these sizes are the most likely to be deposited and build a resistance on the membrane surface.

5.3.2 Particle Size Distribution of Kaolin

A synthetic inorganic water with particles alone (without NOM) was made by adding a clay material to simulate a natural turbid water. The three most common clay types are kaolinite (also called kaolin), sodium montmorillonite, and illite (Sparks 1995). Among these clay materials, kaolin was selected in this research because it is considered to be a non-expansive clay, which reduces problems from the expansion during preparation.

The particle size distribution of kaolin was investigated to understand particle fouling when kaolin was applied to ultrafiltration. Particle size measurements were conducted using Coulter Counter equipment (Coulter Electronics Inc., Hialeah, FL). The Coulter Counter uses the electrical characteristics of an electrolyte solution to size particles in the solution. While a suspension in an electrolytes solution flow through a small opening, a set current is maintained between electrodes on either side of the opening. When a particle moves through the opening, a voltage pulse is created. The magnitude of that pulse is proportional to the volume of the particle (Van Gelder et al. 1999).

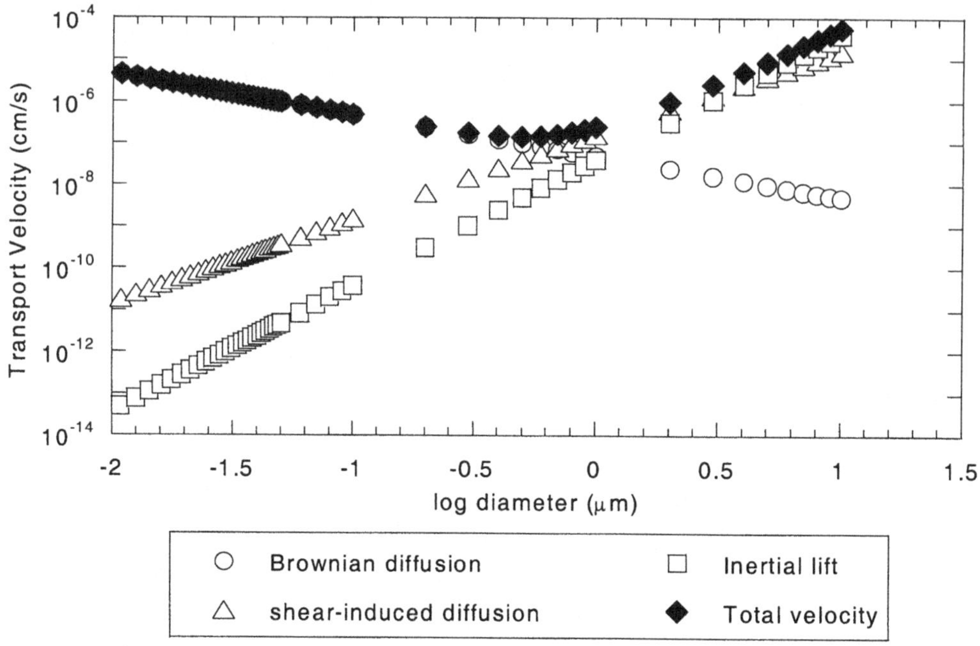

Figure 5.22 Velocities of each mass transport mechanism under the typical operational conditions in this research (crossflow velocity: 10 cm/s and temperature: 20°C)

Figure 5.23 shows the particle number distribution of kaolin. Many particles are present within the size of 0.8 and 2.5 μm diameter. The consideration of mass transport of particles revealed that the particle with the diameter between 0.3 and 1 μm had the minimal back transport velocity. Therefore, some of the particles in the kaolin solution could have a negative effect on membrane fouling because of their small back transport velocity.

5.3.3 Effects of Particle Concentrations on Membrane Fouling

Two different kaolin concentrations were investigated to understand the extent of particle fouling by kaolin resulting in turbidity values of 51 and 720 NTU. The operational conditions and the water qualities are presented in Table 5.16 and Table 5.17, respectively. The operational conditions were properly maintained and the water qualities from two experiments were quite similar except for the expected differences in turbidity of the feed water. A high Ca^{2+} concentration in high turbidity water was the result of ingredients in kaolin itself. Essentially all of the turbidity was removed by the ultrafiltration, as expected.

The flux decline from the two experiments is shown in Figure 5.24. Surprisingly, the flux did not decline much even with the highly turbid water. The experiment with the low turbidity sustained the specific flux at around 100% of the clean water specific flux. Also, only 5% of the clean water specific flux was lost for the high turbidity water at the normalized cumulative production of 150 cm^3/cm^2. Therefore, although a kaolin layer was built up on the membrane surface due to the particles with the minimal back transport velocity, the layer created very little resistance for the water flow.

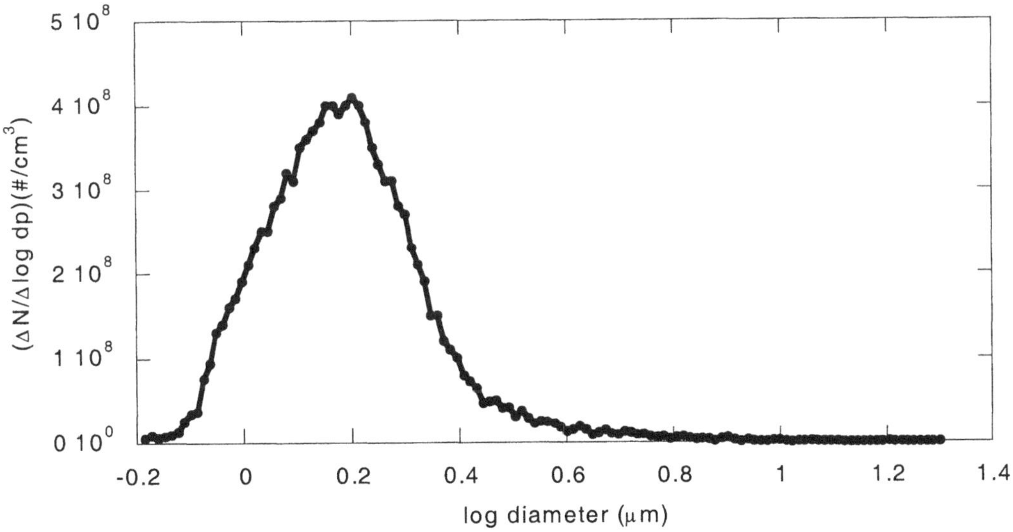

Figure 5.23 Particle number distribution of kaolin used as a surrogate for natural turbid matter

Table 5.16
Operational conditions for experiments with kaolin at different turbidities

Turbidity	TMP (kPa)	Crossflow velocity (cm/s)	Clean waterspecific flux, J_{so} (L/m^2-hr-kPa)
51	87.6 - 90.0	10.0	1.52
720	89.0 - 101.4	10.1	1.32

Table 5.17
Water quality of experiments with kaolin at different turbidities

Experiment	Samples	pH	Turbidity (NTU)	Ca^{2+} (mg/L)	Mg^{2+} (mg/L)
Low turbidity	Feed	8.33	51.1	61.4	18.5
	Filtrate	8.33	0.01	61.2	18.3
High turbidity	Feed	8.38	717.7	70.7	17.4
	Filtrate	8.41	0.01	59.3	17.5

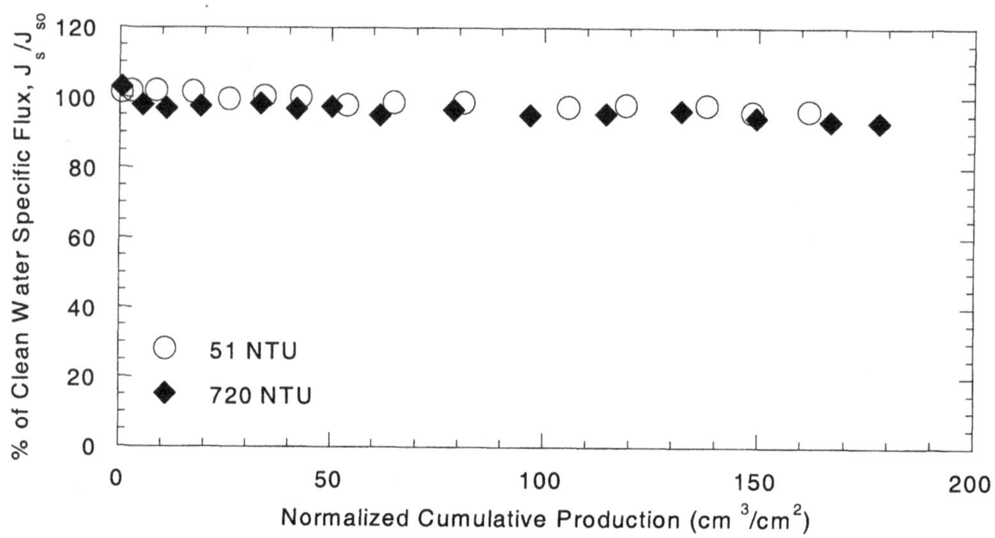

Figure 5.24 Flux decline: effects of turbidity with kaolin

5.4 COMBINATION OF ORGANIC AND PARTICLE FOULING

Little research has been done to investigate combined effects of particle and NOM fouling. NOM fouling has been known to be the biggest foulant in membrane processes. However, particle fouling has been neglected based on the results of fouling by particles only with inorganic properties. As was observed in the previous section, the clay material exhibited little flux decline even with the very high turbidity. However, NOM is likely to be adsorbed on existing particles in natural water. In the presence of NOM, the characteristics of particles such as their stability ("stickiness") and surface charge will be affected; thus, the fouling by particles with adsorbed NOM is expected to be different from inorganic particles, *i.e.*, those with no adsorbed NOM. Therefore, the simple organic components, dextran and alginic acid, were used to evaluate effects of both organic matter and particles on membrane fouling.

5.4.1. Combined Fouling: Dextran with Kaolin

Effects of DOC Concentrations on the Combined Fouling by Dextran and Kaolin

Dextran was used at two concentrations (4 and 20 mg/L C) to investigate the combined fouling of organic matter and kaolin. The turbidity was aimed to be higher than 500 NTU since even the higher turbidity water, i.e., 720 NTU, showed little flux decline without organic matter. The operational conditions of the experiments with kaolin and dextran are shown in Table 5.18. The TMPs and crossflow velocities were reasonably consistent during all operations except the experiment with kaolin and dextran at 20 mg/L C. In that experiment, the TMP and the crossflow velocity were slightly higher than expected. The higher TMP might induce more fouling due to the greater advection flow to the membrane surface but the faster crossflow velocity might help to improve the back transport of foulants, thus compensating for the effect of the high TMP.

The water quality from the experiments with kaolin and dextran are presented in Table 5.19. Since no softening was applied, the water quality did not change much during the experiment except for the turbidity and DOC. As expected, ultrafiltration showed a remarkable removal of turbidity, i.e., almost 99.999% removal of clay at the high turbidity of 734 NTU. The DOC removal at 4 mg/L C solutions was 23.8% with kaolin and 50% without kaolin. At the high concentration of DOC, i.e., 20 mg/L C, the removal was 7.6% with kaolin and 11.8% without kaolin. As seen earlier, the DOC removal decreased at the high DOC concentration regardless of the presence of kaolin. This result implies that a fouling layer of accumulated organic matter is forced down by applied TMP during operation and this process helps the organic matter to pass through membrane. The interesting results were that the DOC removal by ultrafiltration decreased with presence of kaolin, which was not expected. Definitely, dextran does not adsorb well onto kaolin since dextran is relatively hydrophilic. However, kaolin seemed to somehow help more dextran pass through the membrane.

The reductions in flux from the experiments with kaolin and the raw dextran at two concentrations, 4 and 20 mg/L C, are shown in Figure 5.25. The experiments were performed without softening. To compare effects of the presence of kaolin, the results without kaolin are shown as well. Virtually no change in flux was observed with kaolin. It seems that kaolin plays little role in either reducing or increasing membrane fouling by organic matter (i.e., dextran). This result occurred because kaolin has no adsorption capability for dextran, unlike other adsorbents used in membrane processes. Several researchers indicated that adsorbents such as powdered activated carbon and iron oxide helped to improve fouling by adsorbing NOM onto the adsorbent surface before it accumulated and fouled the membrane surface (Laine, Clark, and Mallevialle 1990; Chang 1996).

Table 5.18
Operational conditions for experiments with kaolin and dextran at 4 and 20 mg/L C

Designated DOC (mg/L CaO)	Kaolin	TMP (kPa)	Crossflow velocity (cm/s)	Clean water specific flux, J_{so} (L/m^2-hr-kPa)
4	Yes	89.7 - 96.6	10.1	1.54
4	No	90.3 - 95.2	10.1	1.95
20	Yes	94.5 - 101.4	11.0	1.56
20	No	88.3 - 91.0	10.2	1.56

Table 5.19
Water quality of experiments with kaolin and dextran at 4 mg/L C

Designed DOC	Kaolin	Sample	pH	Turbidity (NTU)	Ca^{2+} (mg/L)	Mg^{2+} (mg/L)	DOC (mg/L)
Low	Yes	Feed Filtrate	8.34 8.38	560.3 0.02	45.4 46.5	16.7 16.9	4.2 3.2
Low	No	Feed Filtrate	8.26 8.25	NA	48.2 45.0	16.9 15.6	5.4 2.7
High	Yes	Feed Filtrate	8.37 8.44	733.7 0.01	52.7 52.5	16.5 16.7	22.3 20.6
High	No	Feed Filtrate	8.41 8.45	NA	59.6 59.8	17.9 17.6	22.0 19.4

Effects of Softening on the Combined Fouling with Kaolin and Dextran

The synthetic organic water with kaolin and dextran was softened at 170 mg/L CaO to investigate effects of softening on membrane fouling. The operational conditions are presented in Table 5.20 along with the experiments conducted without kaolin addition for comparison. The TMP and the crossflow velocities were stable and the clean water specific fluxes fluctuated slightly among the experiments. The water quality of these experiments is summarized in Table 5.21. The enhanced softening (lime dose of 170 mg/L CaO) of the synthetic water with kaolin and dextran improved the water quality of the feed waters for every measurement, *i.e.*, turbidity, Ca, Mg, and DOC. The removal of DOC with the synthetic organic water with kaolin was similar to that without kaolin but the removals of Ca and Mg with kaolin were slightly less than that without kaolin. For instance, the Ca removal was 59% with kaolin and 64% without kaolin and the Mg removal was 70% with kaolin and 82% without kaolin. The reduced removal of Ca and Mg with kaolin might indicate that the presence of kaolin could mildly hinder the precipitation of $CaCO_3$ and $Mg(OH)_2$.

The effects of softening of the synthetic waters with kaolin and dextran on membrane fouling were investigated with the flux decline as shown in Figure 5.26. The results from the experiments without kaolin are also shown to compare the flux decline behaviors. Again, the results clearly show that the presence of kaolin did nothing to change the degree of membrane fouling.

Cleaning Efficiency on the Combined Fouling by Kaolin and Dextran

Results of the efficiency of the various cleaning methods for membranes fouled with kaolin and dextran are shown in Figure 5.27 along with the results from the experiments with dextran only. At the lime dose of 170 mg/L CaO, only a small decline (approximately 10%) of flux was measured. Therefore, it was difficult to evaluate the efficiencies of three cleaning methods within this small percentage of flux. With the raw synthetic waters (without softening), the overall cleaning efficiency was better when kaolin was added in the solution. Almost 99% of the clean water specific flux was recovered in the experiment with kaolin and dextran while approximately 94% of the clean water specific flux was recovered in the experiments only with dextran solution. Also, the caustic wash significantly increased the clean water specific flux in the experiment with kaolin and the raw dextran. Therefore, kaolin addition helped to improve the cleaning efficiency of caustic wash when no softening was applied. However, no clear trends can be seen in the efficiency of the cleaning methods when the enhanced softening was applied.

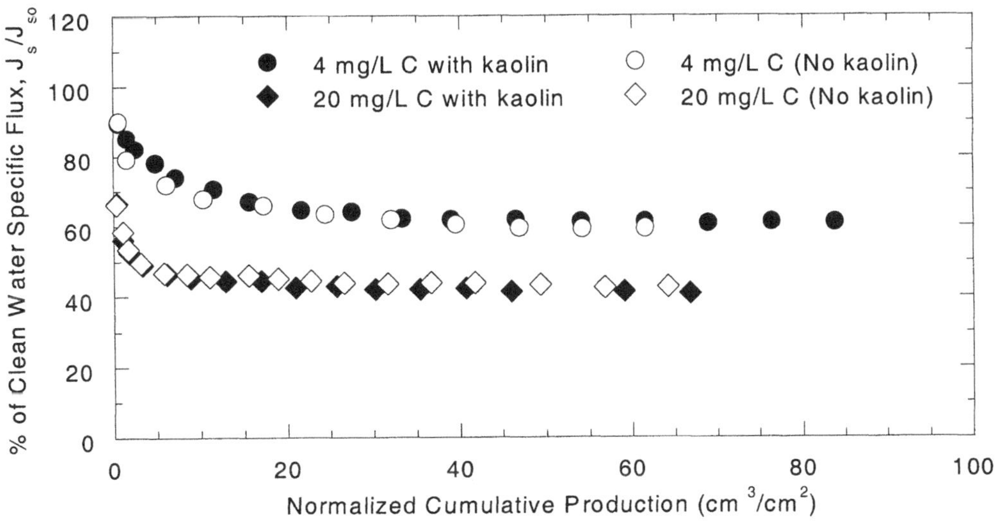

Figure 5.25 Flux decline: effects of DOC concentrations on the combined fouling using kaolin and dextran

Table 5.20
Operational conditions for experiments with kaolin and dextran at two lime doses (4 mg/L C)

Lime (mg/L CaO)	Kaolin	TMP (kPa)	Crossflow velocity (cm/s)	Clean water specific flux, J_{so} (L/m^2-hr-kPa)
0	Yes	89.7-96.6	10.1	1.54
170	Yes	89.7-91.0	10.2	1.88
0	No	90.3-95.2	10.1	1.95
170	No	91.7-95.9	10.1	1.93

Table 5.21
Water quality of experiments with kaolin and dextran at two lime doses (4 mg/L C)

Lime (mg/L CaO)	Kaolin	Sample	pH	Turbidity (NTU)	Ca^{2+} (mg/L)	Mg^{2+} (mg/L)	DOC (mg/L)
0	Yes	Feed	8.34	560	45.4	16.7	4.2
		Filtrate	8.38	0.017	46.5	16.9	3.2
170	Yes	Raw	8.48	649	64.8	17.5	3.6
		Softened	11.23	51.2	26.3	5.2	1.4
		Feed	10.88	2.37	10.4	3.6	NA
		Filtrate	10.92	0.019	8.7	3.6	0.8
0	No	Feed	8.26	NA	59.2	16.9	5.4
		Filtrate	8.25		45.0	15.6	2.7
170	No	Raw	8.06*	NA	56.9	16.6	4.6
		Softened	10.83*	27.2	20.5	3.0	1.9
		Feed	10.56*	4.2	11.1	2.2	1.8
		Filtrate	10.52*	0.013	7.8	1.9	0.7

* Erroneous calibration: low by approximately 0.3-0.4 unit

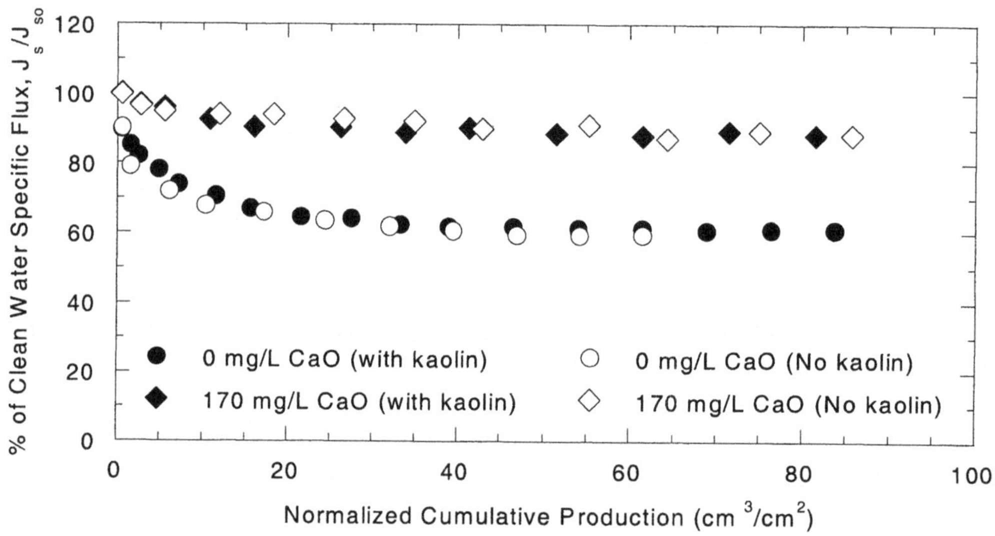

Figure 5.26 Flux decline: effects of softening on the combined fouling using kaolin and dextran (4 mg/L C)

SEM of Membrane Surfaces Fouled by Kaolin and Dextran

The SEM images were taken to investigate directly the surfaces of membranes fouled with kaolin and the raw dextran. As shown in Figure 5.28, the images show a lot of solid deposition (i.e., kaolin) and some fibril shapes (i.e., dextran). The dextran seems to be deposited between the kaolin and the membrane to attach them together. The images with kaolin and dextran showed some similarities and differences from those only with the raw dextran (as shown in Figure 5.11). Image (a) shows a place with many deposits just like the image from Figure 5.11 (b). The image in Figure 5.28 (a) looks like a big lump of deposition that is a mixture of solids (kaolin) and organic matter, while the image in Figure 5.11 (b) seems to be an accumulation of organic matter alone. Image 5.28 (b) shows that kaolin is widely spread all around the membrane area and has many particle sizes of smaller than 2 μm. These results for the sizes of kaolin were consistent with those from the particle size distribution measurement by the Coulter Counter, which indicated the highest concentration of particles between 0.8 and 1.8 μm in diameter. Also, the images in Figure 5.28 (c) and (d) show the magnified raw dextran with kaolin. It seems that kaolin is trapped within web-like knots of dextran, or kaolin is deposited on the top of dextran that is already attached to the membrane surface.

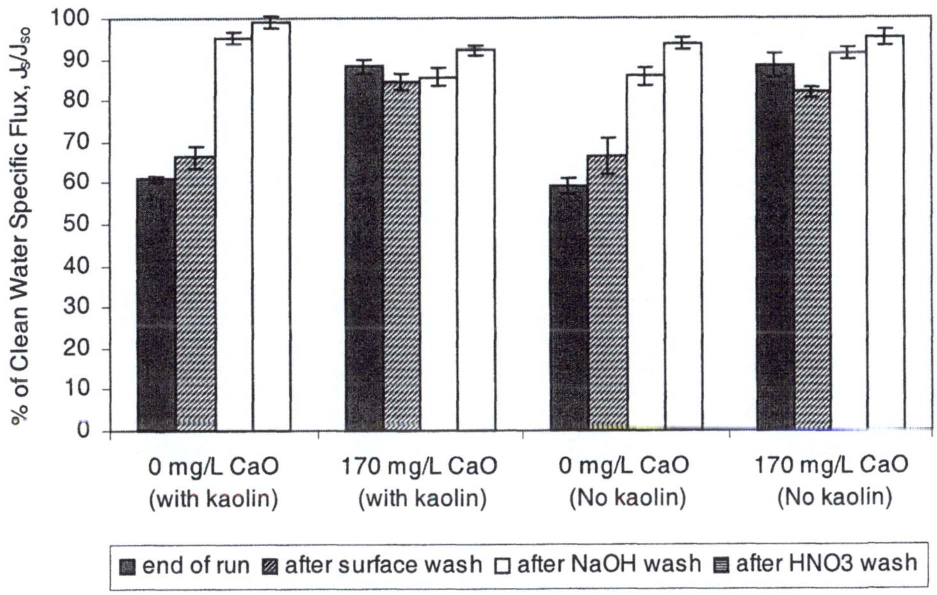

Figure 5.27 Membrane cleaning: synthetic water with dextran and kaolin at two lime doses (4 mg/L C)

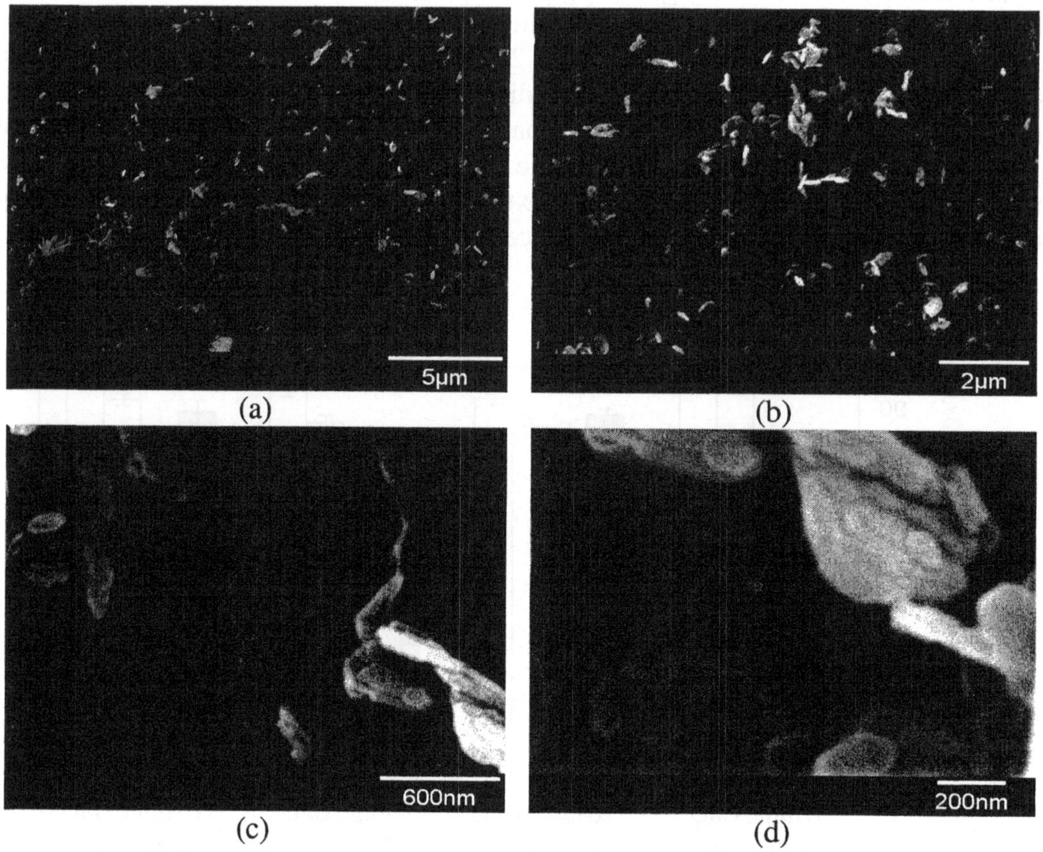

Figure 5.28 SEM images of the membrane fouled by raw synthetic water with dextran and kaolin ([a] 5,000x; [b] 10,000x; [c] 50,000x; and [d] 100,000x resolution)

5.5 MEMBRANE FOULING BY NATURAL WATERS

5.5.1 Effects of Raw Water Characteristic on Fouling

Two natural water sources, Lake Austin water (Austin, TX) and the Missouri River water (St. Louis, MO) were selected for study in this research. These two waters have quite similar water characteristics for inorganic and organic constituents except for turbidity; the turbidity is 1.8 NTU for Lake Austin water and 314 NTU for the Missouri River water. The decline of specific flux was obtained to examine the effect of dramatic differences in particle concentrations from untreated Lake Austin water and the Missouri River water using approximately 10 cm/s of crossflow velocity. The results of flux decline are shown in Figure 5.29. The Missouri River water revealed a significant flux decline with approximately 70% reduction of the clean water specific flux after a normalized cumulative water production of 90 cm^3/cm^2, whereas Lake Austin water showed a relatively small flux decline (*i.e.*, only 30% of reduction of the clean water specific flux); these results support the hypothesis that particle fouling affects membrane performance.

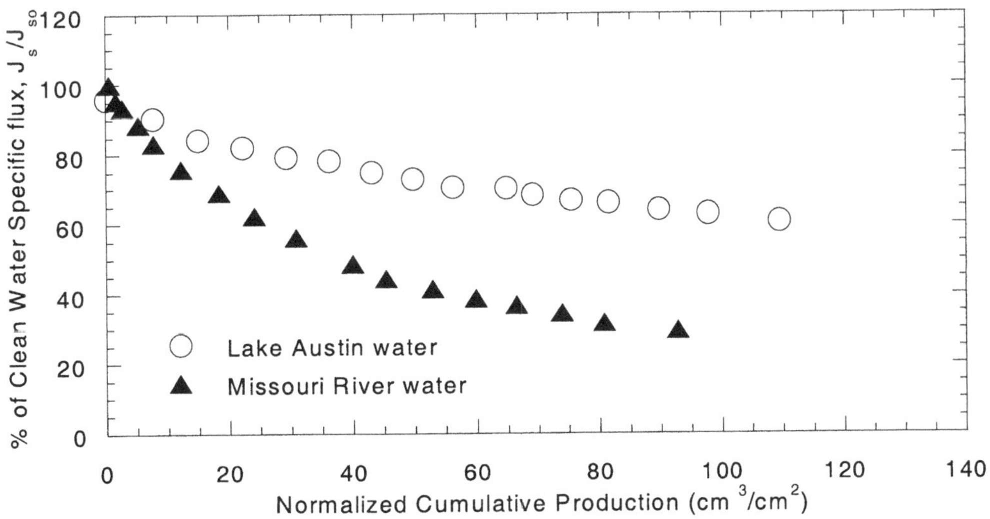

Figure 5.29 Flux decline: two natural water sources

Table 5.22
Operational conditions of experiments for two natural water sources

Source	Membrane	TMP (kPa)	Crossflow velocity (cm/s)	Clean water specific flux, J_{so} (L/m^2-hr-kPa)
Lake Austin	Polysulfone	90.3-95.2	10.3	1.57
Missouri River	Polysulfone	89.0-93.8	10.4	2.50

The operational conditions of these tests are summarized in Table 5.22. Transmembrane pressure (TMP) was supposed to be maintained at 90 kPa but it fluctuated slightly during the operations. The crossflow velocity was controlled to approximately 10 cm/s. The clean water specific flux is the measured flux using distilled/deionized water for one hour after thorough a wetting and cleaning procedure. The clean water specific fluxes varied considerably among the different membrane sheets, which demonstrates the reason that the normalization of the results is necessary.

Table 5.23 shows the water quality of the two natural water sources during these experiments. Again, the feed water characteristics were similar except for the turbidity. The ultrafiltration showed an excellent turbidity removal, i.e., 99.99% of removal from the Missouri River water. Considering the high turbidity removal in the Missouri River water, the more rapid flux decline is likely related to an increase in the hydraulic resistance from the accumulated solids on the membrane surface. Also, the ultrafiltration removed approximately 12% of the organic carbon (DOC) from Lake Austin water and 19% from the Missouri River water.

The levels of softening for the two water sources were chosen based on the softening and enhanced softening performance, as reported in Chapter 4. Lake Austin water was used as the primary source in this research, and so the synthetic waters with inorganic and/or organic constituents were simulated as Lake Austin water. In this section, results from thorough investigations of the two natural water sources on membrane fouling are presented. Results from the experiments with Missouri River water illustrate effects of turbidity by comparing to Lake Austin water. In addition, experiments with the two natural waters are compared to experiments with synthetic organic (dextran or alginic acid) water and synthetic water with kaolin and dextran.

5.5.2 Investigation of Lake Austin water on membrane fouling

Effects of Extent of Softening on Fouling by Lake Austin Water

The results of softening and NOM removal at various lime doses for Lake Austin water were presented in Chapter 4. Based on the extent of softening by lime softening alone, three lime doses were selected: 125, 170, and 230 mg/L as the standard softening, the enhanced softening, and the Mg softening conditions. Each condition represents different water quality with respect to inorganic and organic matter.

The effect of the levels of softening on fouling in the subsequent ultrafiltration was examined using lime softening alone at the three lime doses. The operational conditions and the water quality are shown in Chapter 4, where various softening scenarios were examined. In general, the crossflow velocity and TMPs were maintained within the proper ranges. The clean water specific flux substantially fluctuated from sheet to sheet as seen in other experiments.

The flux decline as a function of the normalized cumulative production is shown in Figure 5.30. The fluxes decreased with similar patterns regardless of the lime dose, except for the 230 mg/L dose. The flux decline of the softened water at 125 mg/L lime dose was almost identical to that of raw Lake Austin water, despite differences in pH and Ca^{+2} concentration and some reduction in NOM. At the lime dose of 170 mg/L, the percent of the clean water specific flux is greater than that of raw water during the whole operation. For instance, after 120 cm^3/cm^2 of normalized cumulative production, the percent of the clean water specific flux was 68.0% for the softened water at 170 mg/L CaO of lime and 60.6% for the raw Lake Austin water. However, the improved clean water specific flux was quite small, i.e., less than 10%.

At the highest dose (230 mg/L as CaO), the decline was quite gradual, so at the end of the run, the reduction of the clean water specific flux was only 15%. The very gradual decline of specific flux in this high lime dose experiment might stem from the lower NOM concentration that was achieved through softening. The DOC in the UF feed was 1.6 mg/L in this experiment as compared to a minimum of 2.6 mg/L at the other conditions. Also, the higher pH might have increased the repulsive interaction between the membrane surface and NOM, which would help to reduce NOM fouling. If NOM is the primary cause of fouling, the fraction of NOM that is removed only at the Mg softening condition apparently plays a significant role in fouling of ultrafiltration.

Table 5.23
Water quality of experiments for two natural water sources

Source	Sample	pH	Turbidity (NTU)	Ca^{2+} (mg/L)	Mg^{2+} (mg/L)	DOC (mg/L)	SUVA (L/m-mg)
Lake Austin	Feed	8.29	2.0	66.2	19.9	3.3	3.0
	Filtrate	8.33	0.2	66.1	20.0	2.9	2.8
Missouri River	Feed	8.31	297	53.3	16.7	3.2	3.0
	Filtrate	8.29	0.02	46.9	14.7	2.6	3.4

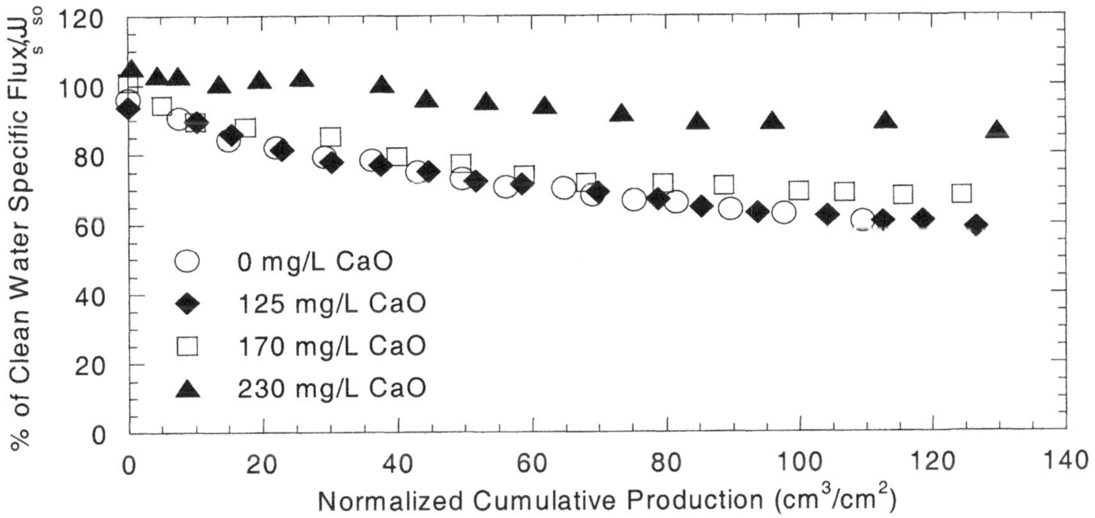

Figure 5.30 Flux decline: softening Lake Austin water at different lime doses (lime softening alone)

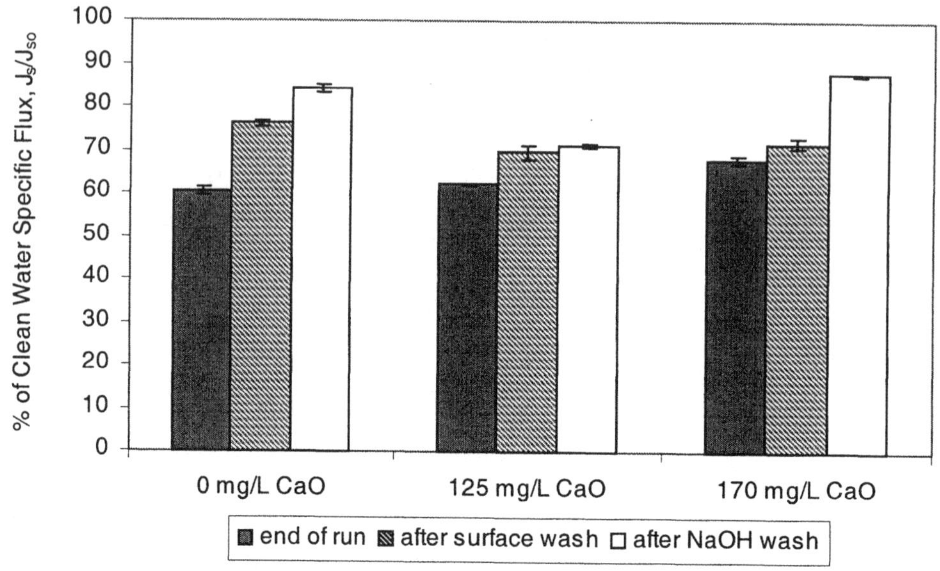

Figure 5.31 Membrane cleaning: Lake Austin water at different lime doses

Efficiency of Cleaning Methods on Lake Austin Water

The clean water specific flux was measured to evaluate the recovery obtained by the cleaning procedures. The experiments were performed relatively early in the research. At this time, it was hypothesized that NOM fouling would be great but inorganic fouling would be negligible. Hence, the cleaning procedures only included the surface wash with distilled/deionized water and the NaOH wash. The cleaning efficiency was analyzed for the raw water and the softened water at the lime doses of both 125 and 170 mg/L CaO. The lost flux at the lime dose of 230 mg/L CaO was too small to evaluate the cleaning efficiency adequately.

The results of each cleaning are shown in Figure 5.31. For the raw water, both cleaning methods increased the flux measurably, suggesting that the cake was only loosely attached to the membrane and contained both particulate and organic materials. For the lime doses, the surface wash had negligible effect. Particulate matter in these waters would be predominantly precipitated solids ($CaCO_3$ or $Mg(OH)_2$) rather than the various debris that might have been in the raw water, and these solids were not dislodged much by the surface wash. The sodium hydroxide was ineffective at the lime dose of 125 mg/L CaO, but was successful at the lime dose of 170 mg/L CaO. The inefficiency of the caustic wash at the standard softening condition indicated that further cleaning with an acidic wash was necessary.

Molecular Weight Distribution in Softening of Lake Austin Water

The molecular weight distributions were analyzed with size exclusion chromatography (SEC) for the raw and the softened Lake Austin water at the lime dose of 170 mg/L CaO. As shown in Figure 5.32, the chromatograms illustrate two peaks in the raw and the softened Lake Austin water: one at a molecular weight of approximately 1400 Da and another at MW < 300 Da.

As noted earlier, the ordinate of the chromatograms shows the relative DOC concentration since interest is in the changes of molecular weight distribution, not in the absolute changes in the DOC concentration. The two curves shown are very similar, indicating that softening at 170 mg/L CaO had very little effect on the makeup of NOM fed to the membranes.

Previously, the research by Laine, Clark, and Mallevialle (1990) identified two peaks at approximately 1,000 Da and larger than 100,000 Da in a natural water (Lake Decatuer, Ill.). The authors compared the chromatogram before and after coagulation-flocculation and concluded that the process was relatively more effective in removing the high-MW compounds (*i.e.*, MW > 10,000 Da). In research by Bruchet, Rousseau, and Mallevialle (1990), the authors monitored the changes in the molecular weight distributions of raw and clarified waters. The clarification process was performed on the Seine River water by slow sand filters or direct filters. The water had three distinctive peaks at 8500 Da, 1300 Da, and 500 Da. The clarification process mainly removed DOC from the higher MW fractions.

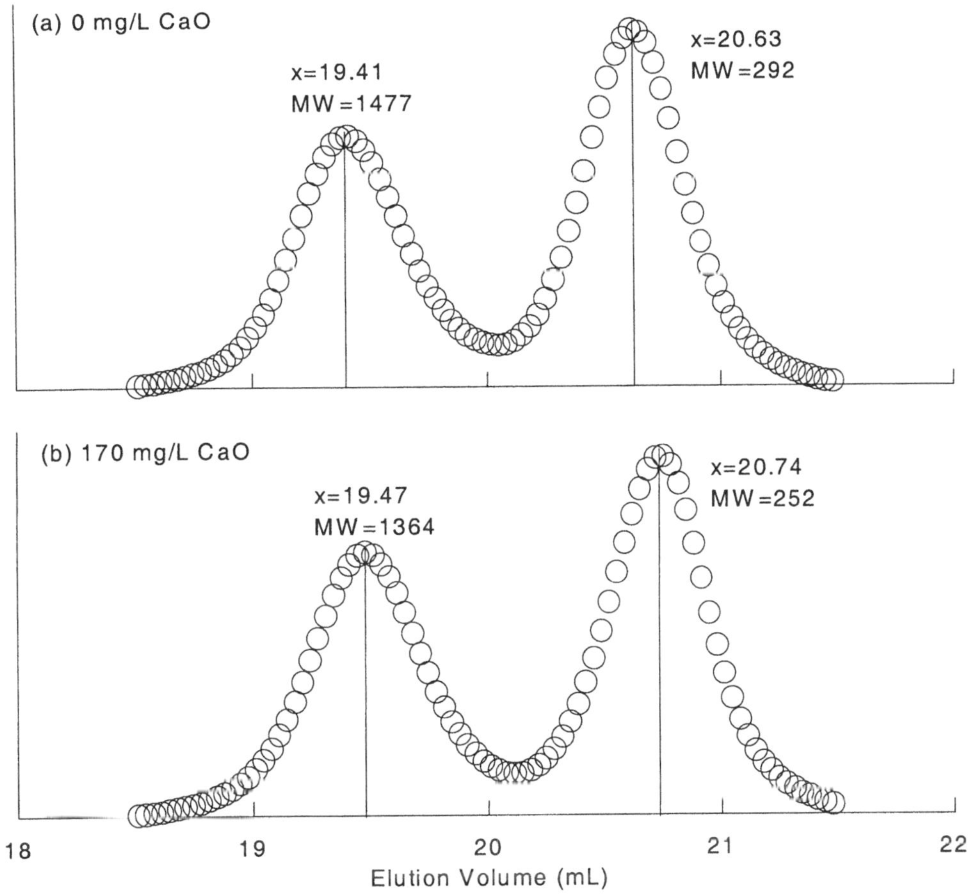

Figure 5.32 Changes in molecular weight distribution by softening Lake Austin water

Compared to previous research, the Lake Austin water seems to be missing high molecular weight organic matter. The absence of the high molecular weight fraction might explain why Lake Austin water has a small hydrophobic portion of DOC (Thompson et al. 1997) since the high molecular weight fraction is generally more hydrophobic than the low MW fraction. Also, the absence of much high molecular weight material might explain why the enhanced softening, i.e., 170 mg/L CaO, had little effect on the reduction of flux decline (Figure 5.30). Although the DOC concentration was reduced at the lime dose of 170 mg/L, the removal of DOC was primarily in the relatively low molecular weight fraction, which might have little effect on fouling.

Hydrophobic DOC Fraction in Softening of Lake Austin Water

In general, the hydrophobic fraction of organic matter is known to be responsible for the most of flux decline. To investigate changes of hydrophobic DOC in the softening process, the DOC in Lake Austin water after softening was fractionated by its hydrophobicity using XAD-8 resin. The results are summarized in Table 5.24. In these measurements, the DOC in the raw water was high compared to most of our measurements and reported above for the raw water characteristics. However, the percent of hydrophobic DOC in the different measurements was approximately the same, i.e., 46% in this experiment and 44% in the experiment for the raw water characteristics.

The hydrophobic DOC fraction generally decreased as the degree of softening increased, although it was the same at the lime doses of 125 and 170 mg/L CaO. The decrease in the hydrophobic fraction of DOC is consistent with the results of the specific UV absorbance (SUVA) at 254 from the experiment to determine the enhanced softening performance in the Chapter 4.

The most dramatic reduction in the hydrophobic fraction occurred at the lime dose of 230 mg/L CaO. The hydrophobic DOC fraction dropped from 46% in the raw Lake Austin water to only 19%. The SUVA values had a similar trend, *i.e.*, the decreases of the SUVA were 16, 20, and 41% as the lime doses increased (as presented in Table 4.2). These results explain why the flux reduction was very substantial at the lime dose of 230 mg/L CaO. The $Mg(OH)_2$ precipitation apparently had a significant role in removing the hydrophobic fraction of NOM, which, in this water, has relatively low molecular weight and a strong tendency to foul the membrane.

Table 5.24
Hydrophobic fraction of NOM during softening Lake Austin water

Lime (mg/L CaO)	Soda ash (mg/L CaO)	DOC (mg/L C)	Hydrophilic DOC (mg/L C)	Hydrophobic DOC (mg/L C)	Fraction of hydrophobic DOC (%)
0	-	5.9	3.2	2.7	46
125	-	4.6	2.8	1.7	38
170	45	3.9	2.4	1.5	38
230	105	3.6	2.9	0.7	19

SEM of Membrane Surfaces Fouled by Lake Austin Water

The SEM images from the membrane fouled with the softened Lake Austin water at the lime dose of 230 mg/L CaO and the soda ash dose of 105 mg/L CaO are presented in Figure 5.33. The images were taken at relatively low resolutions, i.e., 1-2k and 10-20k, since the deposits were relatively large. The images show some isolated inorganic particles and their aggregates as well as inorganic particles associated within a fibril network of organic matter. These deposits were relatively widespread over the whole area of the membrane surface. It seems that organic matter acts like a glue to stick inorganic particles together. The inorganic particles within organic matter in Lake Austin water were well defined, as rhombus shapes, the typical morphology of calcite, were detected. Compared to the images from the Lake Austin water, the SEM images from the experiment with Dextran60 solution at the same condition had more sheet-like deposits and less distinguished edges (Figure 5.13).

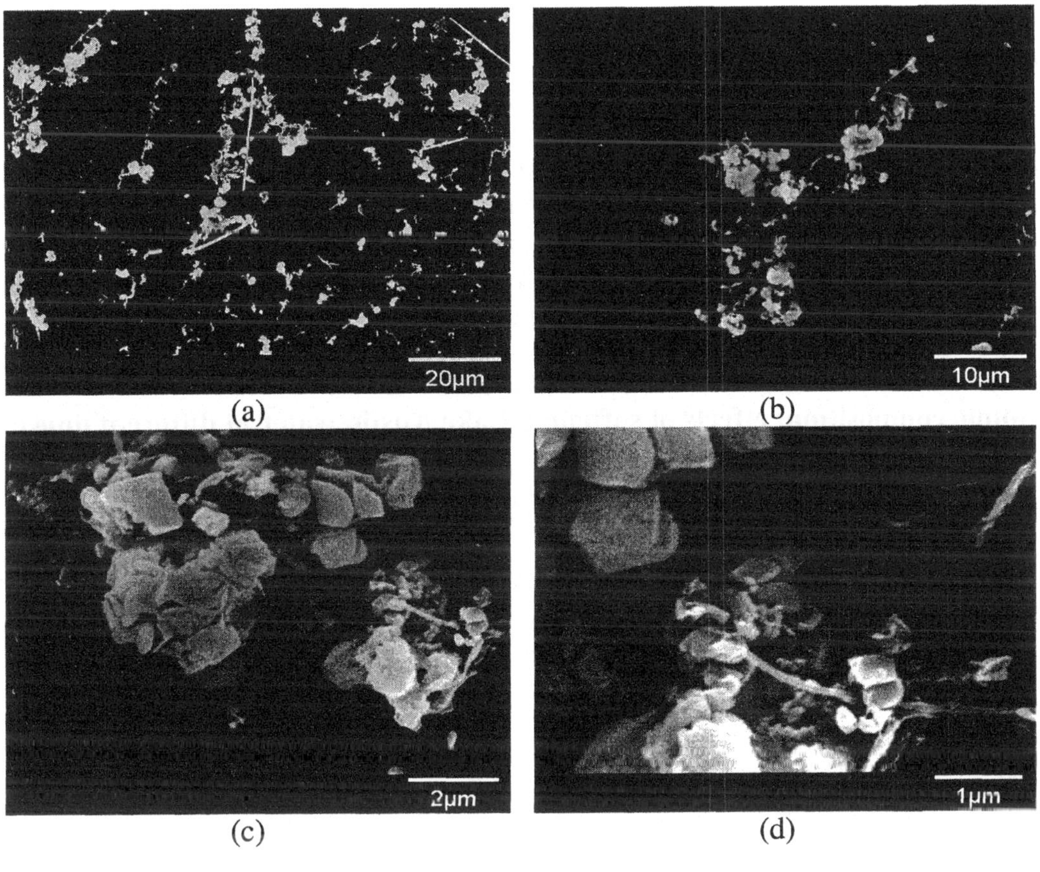

Figure 5.33 SEM images of the membrane fouled by the Lake Austin water softened at 230 mg/L of lime ([a] 1,000x; [b] 2,000x; [c] 10,000x; and [d] 20,000x resolution)

XPS of Membrane Surfaces Fouled by Lake Austin Water

The atomic composition of the top layer of the membranes fouled with Lake Austin water was measured with X-ray photoelectron spectroscopy (XPS). The composition of the atoms that were dominant at the surface of the membrane is presented in Table 5.25. The composition of Si(2p) was included since the research by Ralls (1999) and Smith (2001) indicated that substantial amounts of silicate were detected in Lake Austin water. In that work, Si was shown to have significant role in the precipitation of magnesium, apparently forming some magnesium silicate solid.

The composition of C(1s) in the raw water and the softened water at the lime dose of 230 mg/L CaO is smaller than in the other waters. The relatively significant composition of Mg and Si in the raw water induced the decrease of the relative concentration of C(1s). For the highest lime dose, the increased amount of O(1s) as well as inorganics were the cause of the decreased C(1s) composition. The amount of Ca(2p) was negligible in the membranes fouled with the softened Lake Austin water at the lime dose of 125 and 170 mg/L CaO. The increase of lime dose to 230 mg/L CaO increased the Ca and Mg concentration. The increase in Ca(2p) indicated that the added lime increased the Ca^{2+} concentration in the solution since no soda ash was added.

Perhaps the most important result was the deposition of Mg and Si on the membrane for the sample with no lime (*i.e.*, raw Lake Austin water). Evidence from the softening study (Smith 2001) has suggested that a precipitate that includes Ca^{2+}, Mg^{2+}, and silica is formed in Lake Austin water, and these results supported that conclusion. Several different solids with the constituents are thermodynamically possible, and it appears that tremolite ($2CaO{\cdot}5MgO{\cdot}8SiO_2$) could be the most likely. The XPS results reported here confirm the close association of Mg and Si in the precipitation that apparently occurs in the water.

Table 5.25
Atomic composition: effects of softening Lake Austin water at different lime doses

Lime dose (mg/L CaO)	C(1s) (%)	O(1s) (%)	S(2p) (%)	Ca(2p) (%)	Mg(2s) (%)	Si(2p) (%)
0	66.2	21.6	3.7	0.9	4.0	3.8
125	72.4	21.7	4.4	0.4	0.0	1.2
170*	71.4	22.3	4.9	0.0	0.0	1.4
230*	65.4	27.5	3.1	1.3	1.1	1.7

* No soda ash was added

The increase in Mg(2s) at the lime dose of 230 mg/L CaO certainly stemmed from the $Mg(OH)_2$ precipitation, which was not completely removed in the settling process. The atomic composition from the membrane fouled with the raw Lake Austin water verifies that Mg and Si were present in the raw water and deposited on the membrane surface in substantial amounts. In the softened waters, the Si and Mg were greatest at the lime dose of 230 mg/L, which is consistent with the results from Ralls (1999) and Smith (2001) that Si was precipitated with Mg under the Mg softening condition.

The binding energy of each atom is summarized in Table 5.26 along with its typical value. As expected, the binding energies of all the elements varied around their typical values. The binding energy of O and S were the characteristics of the elements in polysulfone, i.e., 531.7 eV of oxygen and 167.9 eV of sulfur. The binding energies of Ca and Mg were difficult to detect because of their weak signals from the low concentration on the membrane surfaces. Compared to Ca and Mg, the Si showed relatively big peaks indicating that considerable amounts were present on the surface.

Detailed spectra of C(1s), O(1s), and Si(2p) of membranes fouled with Lake Austin water at different lime doses are shown in Figure 5.34. Because the amount of Si(2p) was quite minor, the spectra exhibited low intensities with relatively significant background noise compare to that of C(1s) and O(1s). The spectra of C(1s) from the membranes fouled with Lake Austin water (Figure 5.34 (a)) showed relatively long tailed shapes regardless of the lime dose, compared to the clean membrane. The tail of the peaks from the membranes fouled with Lake Austin water was broader than those fouled with dextran, but narrower than those with alginic acid. The long tailed shapes corresponded to the increased intensity at the binding energy of 286.6 eV, which was ascribed to C – O bond in C(1s). The increased intensity at the binding energy of C – O is consistent with the spectra of O(1s) shown in Figure 5.34 (b). The membranes fouled with Lake Austin water all showed increased intensities at binding energies between 532 and 534 eV, which are approximately in the range of the binding energy of O(1s) due to C – O bond (533.5 eV).

The spectra of O(1s) for the membrane fouled with the softened Lake Austin water at the lime dose of 230 mg/L CaO also showed increases in the intensity of the binding energies between 530 and 532 eV. Considering the binding energies of O(1s) in CaO (531.3 eV), MgO (532.1 eV), and SiO_2 (532.8 eV) are approximately in the range, the increased intensity in this range suggests the influence of inorganic solids. This interpretation is also supported by the increased intensity of spectra of Ca(2p), Mg(2s), and Si(2p) at that lime dose.

Although the spectra showed a high degree of background noise, the spectra of Si(2p) from the membrane fouled with raw Lake Austin water were quite different from those fouled with softened Lake Austin water as shown in Figure 5.34 (c). The spectra from the membrane fouled with the softened Lake Austin water showed that the Si concentrations increased with increasing lime doses.

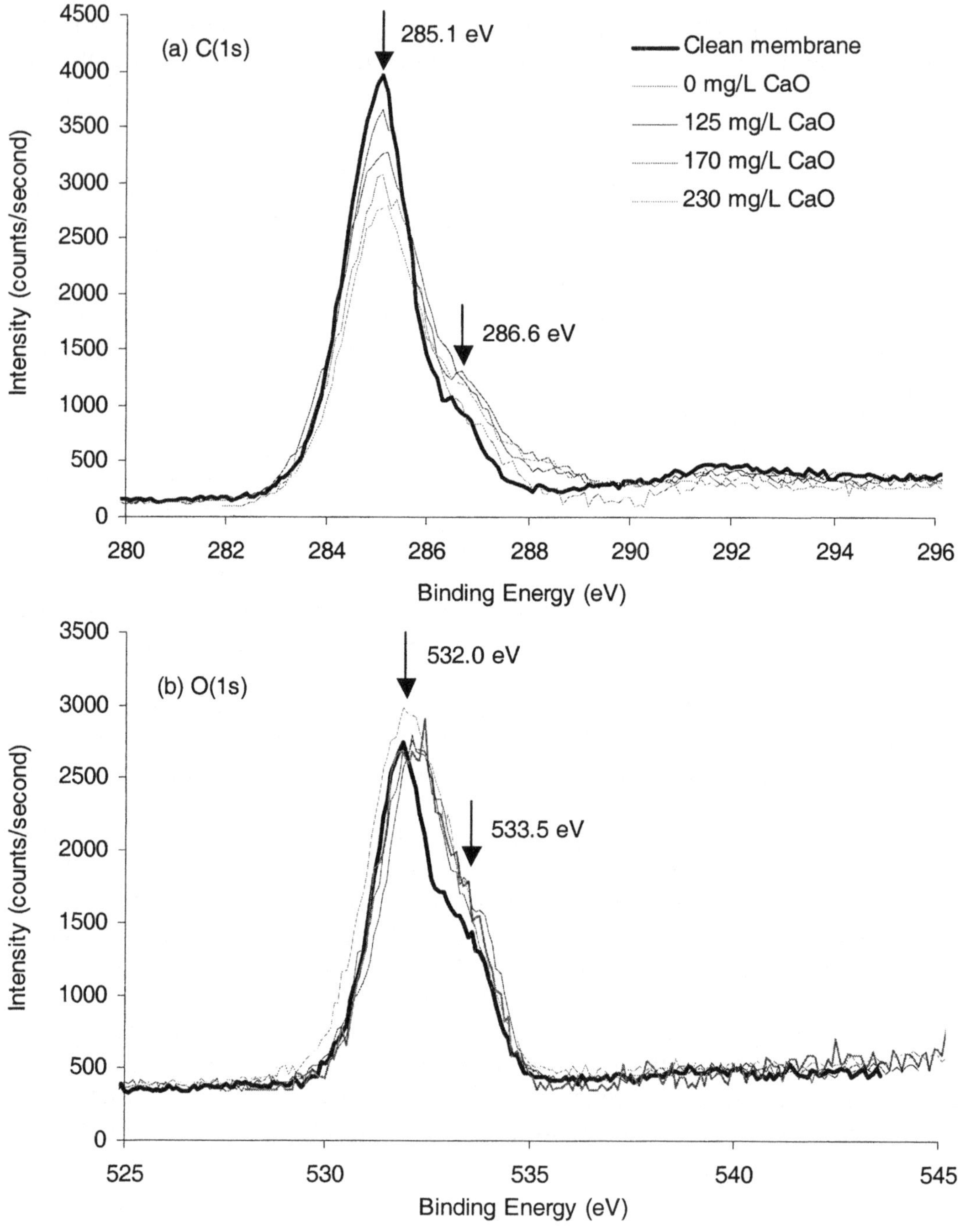

Figure 5.34 Spectra of the membrane fouled with Lake Austin water at different lime doses

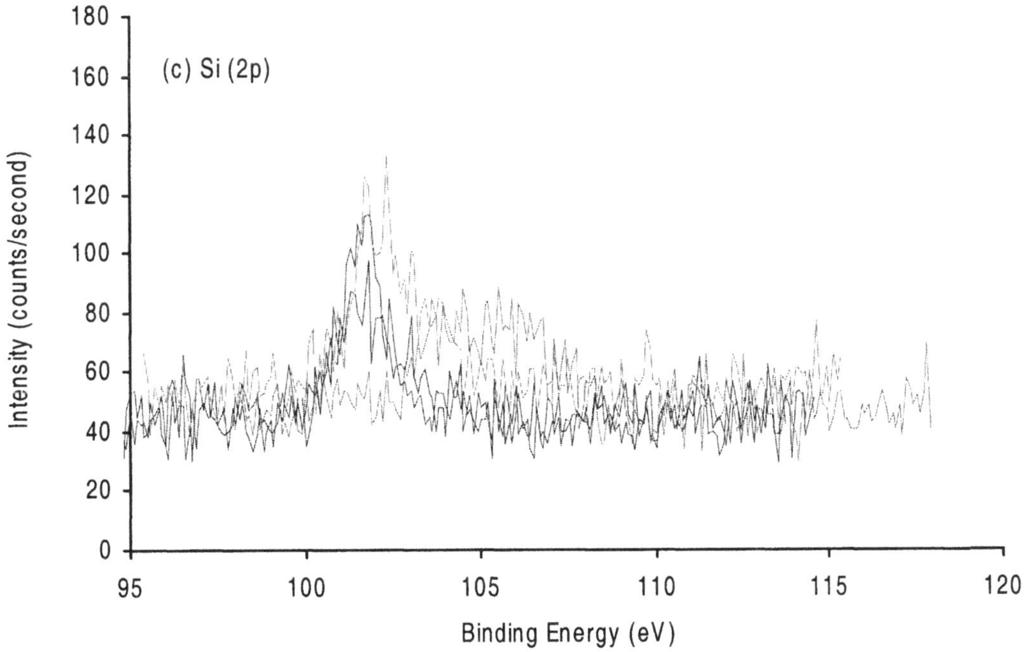

Figure 5.34 Continued

Table 5.26
Binding energies of each atomic composition of the membranes fouled with Lake Austin water

Atom	Survey range	Typical value*	Lime dose (mg/L)			
			0	125	170	230
O(1s)	525-545	531	532.0	532.0	532.1	531.7
S(2p)	158-178	164	168.0	168.2	168.3	168.0
Ca(2p)	342-362	347,351	NA	NA	NA	347.9,351.3
Mg(2s)	84-104	89	93.5	NA	NA	NA
Si(2p)	95-115	99.3	104	101.8	101.8	102.1

*adapted from Moulder J. F. et al. (1992), Handbook of X-ray Photoelectron spectroscopy.
Note: The binding energy of C(1s) fixed at 285.1 eV.

5.5.3 Investigation of the Missouri River Water on Membrane Fouling

The Missouri River water (St. Louis, MO) was selected to emphasize effects of turbidity on membrane fouling. Through softening under various lime doses, two lime doses were selected for the Missouri River water: 90 mg/L CaO for the standard and enhanced softening condition and 165 mg/L CaO for the Mg softening condition. The $Mg(OH)_2$ precipitation started approximately at the lime dose for the optimum Ca removal, so the standard and enhanced softening condition was at the same lime dose. Membrane fouling was investigated with these two lime doses for the Missouri River water.

Effects of Extent of Softening on Fouling by the Missouri River Water

Since the Missouri River water had tremendous amounts of particulate matter, lime softening alone showed ineffective removal of turbidity. Previous research, as well as operation of the plant that treats this water, suggested adding an iron dose of 3 mg/L Fe^{3+} to improve the turbidity removal (Smith 2001). Therefore, the iron was introduced to the Missouri River water during the rapid mixing period. Therefore, three chemicals, lime, soda ash, and Fe were added in that order during the rapid mixing at the high lime dose of 165 mg/L CaO.

The operational conditions are presented in Table 5.27. The TMP varied during the operation within a range of 85 to 99 kPa. The crossflow was slightly higher than the expected value (10 cm/s). The clean water specific flux was quite stable around 2.1 to 2.6 L/m^2-hr-kPa, which was a relatively high flux compared to those in the previous experiments (approximately from 0.7 to 2.1 L/m^2-hr-kPa).

The water quality for the experiments with the Missouri River water is summarized in Table 5.28. As expected, the pH increased as lime doses were increased. The raw water turbidity was in the range of 297 to 405 NTU and was significantly reduced to 6 - 9 NTU by softening with iron addition. The Ca^{2+} and Mg^{2+} removal also increased as lime doses were increased and soda ash was added at the highest dose. The DOC removal by softening increased as lime doses increased: 11% of DOC removal at the lime dose of 90 mg/L CaO and 29% of DOC removal at the lime dose of 165 mg/L CaO. The DOC removal by ultrafiltration varied slightly between 13 and 18%. The SUVA decreased in the softened water as lime doses were increased.

Table 5.27
Operational conditions for experiments with the Missouri River water

Lime (mg/L CaO)	Soda ash (mg/L CaO)	TMP (kPa)	Crossflow velocity (cm/s)	Clean water specific flux, J_{so} (L/m^2-hr-kPa)
0	0	89.0 - 93.8	10.4	2.50
90*	0	84.8 - 98.6	10.3	2.56
165*	75	91.0 - 93.1	11.0	2.12

*Addition of 3 mg/L Fe^{3+} during softening

Table 5.28
Water quality of experiments with the Missouri River water

Lime (mg/L CaO)	Soda ash (mg/L CaO)	Samples	pH	Turbidity (NTU)	Ca^{2+} (mg/L)	Mg^{2+} (mg/L)	DOC (mg/L)	SUVA (L/m-mg)
0	0	Feed	8.31	297	53.3	16.7	3.2	3.01
		Filtrate	8.29	0.019	46.9	14.7	2.6	3.44
90	0	Raw	8.05	325	46.8	14.2	3.7	2.47
		Softened	10.15	9.0	24.9	10.4	3.3	2.44
		Feed	9.47	2.43	23.6	10.2	3.3	NA
		Filtrate	9.53	0.015	23.3	10.0	2.8	2.02
165	75	Raw	7.95	405	60.8	17.4	3.1	3.61
		Softened	11.00	6.0	13.2	1.0	2.2	2.46
		Feed	10.74	3.03	10.4	0.9	2.2	2.56
		Filtrate	10.78	0.026	7.8	0.7	1.8	2.07

*Addition of 3 mg/L Fe^{3+} during softening, # TOC: 3.8 mg/L C

The results of flux decline from the experiments with the raw and the softened Missouri River water are shown in Figure 5.35. The flux decline with the raw Missouri River water was very rapid; therefore, only 30% of the clean water specific flux remained after 90 cm^3/cm^2 of the normalized cumulative production. Softening shows dramatic improvement of the flux decline of the raw Missouri River water. At the lime dose of 90 mg/L CaO, 75% of the clean water specific flux was still maintained after 90 cm^3/cm^2 of the normalized cumulative production. The higher lime dose (the Mg softening condition) yielded only a marginal further improvement of the flux, despite the greater precipitation of $CaCO_3$ and $Mg(OH)_2$, and the increased DOC removal.

The great flux improvement between the raw and softened waters was surprising from two feed waters with trivial differences in DOC concentrations. Note that the DOC concentration was 3.2 mg/L C in the raw water (total organic carbon concentration was reported as 3.8 mg/L C in Table 5. 28) and the DOC concentration was 3.3 mg/L C in the feed water for the experiments at the lime dose of 90 mg/L CaO. The results suggest that the DOC that is preferentially removed in softening has a greater role in fouling of ultrafiltration than the amount of DOC remaining after the softening process. In addition, the removal of turbid matter from the raw water by softening might play a role in improving the flux. The turbidity of the raw water was quite high (297 NTU) compared to the feed water after softening at 90 mg/L CaO (2.43 NTU). The particles in the raw Missouri River might have crucial sizes to foul the membrane, or could have properties that cause them to accumulate and build a fouling layer with significant resistance. The adsorption of NOM on particle surfaces in the natural water could cause the characteristics of the particles to preferentially foul the membrane. Note that this behavior in natural water is different than that observed in the synthetic water; water with both kaolin and dextran yielded the same flux decline as the water with dextran only because kaolin showed little adsorption capability of dextran. In addition, the flux decline from the experiment with the raw Missouri River water was greater, *i.e.*, 30% of the clean water specific flux at 80 cm^3/cm^2 of the normalized cumulative production, than one the synthetic water made with dextran and kaolin, *i.e.*, 60% of the clean water specific flux at the same production. The turbidity of synthetic water with dextran and kaolin were actually higher (560 NTU) than the Missouri River water (297 NTU) although both waters had similar DOC.

Effects of Settling on Fouling by the Missouri River Water

The result from the experiment at the lime dose of 90 mg/L CaO showed a substantial improvement of the flux in ultrafiltration. This improvement could stem simply from the removal of particulate matter from the system; or it could come from changes in particle properties. Softening (including coagulation) changes the particles to bigger flocs, and also changes particle composition from natural particles to only $CaCO_3$ and $Fe(OH)_3$, which are coprecipitated with NOM and flocculated with natural particles. Therefore, the effects of particulate matter in the system were investigated with the water that had been softened water at 90 mg/L CaO but not settled; this water had tremendous amounts of flocs from softening.

The operational conditions are summarized in Table 5.29. The experiments with and without settling at the lime dose of 90 mg/L CaO are shown together for comparison. The crossflow velocity and the clean water specific flux were stable, but the TMP rose dramatically during the experiment without settling because of the high particle load. The water quality achieved in these experiments is presented in Table 5.30. Little difference was shown in the two experiments; pH rose due to the lime addition, and there were small decreases in Mg^{2+}, DOC, and SUVA. The turbidity in the softened water became higher in the softened water (i.e., 785 NTU) than the raw water (i.e., 454 NTU) because of precipitates, mainly $CaCO_3$. Also, total Ca^{2+} was increased to 124 mg/L due to the lime addition and the soluble Ca^{2+} concentration was approximately 22.0 mg/L in the feed water.

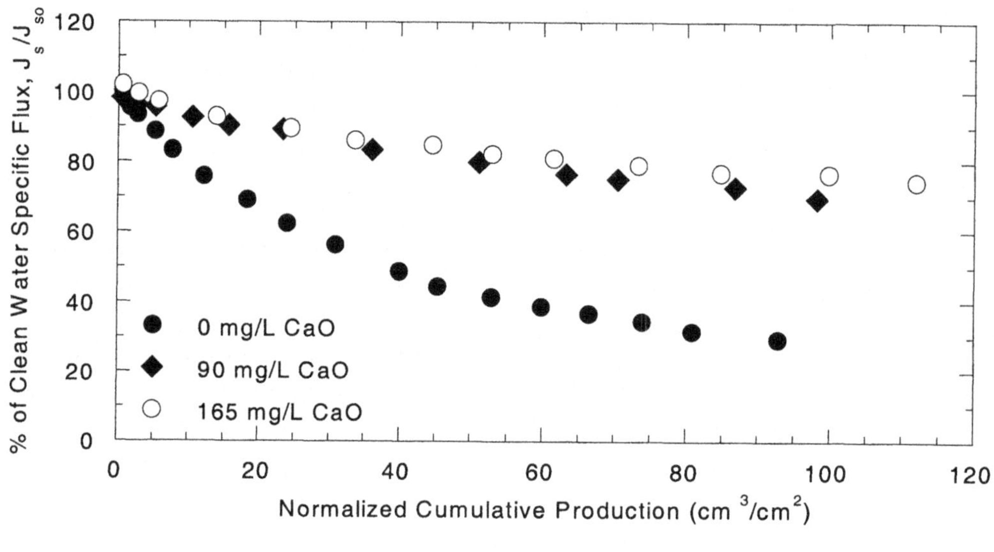

Figure 5.35 Flux decline: effects of softening Missouri River water at different lime doses

Table 5.29
Operational conditions for experiments with the Missouri River water with and without settling

Lime (mg/L CaO)	Settling	TMP (kPa)	Crossflow velocity (cm/s)	Clean water specific flux, J_{so} (L/m^2-hr-kPa)
90	No	90.9-113.8	10.1	2.20
90	Yes	84.8-98.6	10.3	2.56

Table 5.30
Water quality of experiments with the Missouri River water with and without settling

Lime (mg/L CaO)	Settling	Sample	pH	Turbidity (NTU)	Ca^{2+} (mg/L)	Mg^{2+} (mg/L)	DOC (mg/L)	SUVA (L/m-mg)
90	No	Raw	8.10	454	59.0	17.9	2.5	3.28
		Softened	9.90	785	124.0*	18.3	2.3	2.50
		Feed	9.35	795	22.0#	10.9	2.4	2.55
		Filtrate	9.17	0.015	22.6	10.7	2.3	2.47
90	Yes	Raw	8.05	325	46.8	14.2	3.7	2.47
		Softened	10.15	9.0	24.9	10.4	3.3	2.44
		Feed	9.47	2.43	23.6	10.2	3.3	NA
		Filtrate	9.53	0.015	23.3	10.0	2.8	2.02

* soluble Ca^{2+} concentration was 23.7, # reported as soluble Ca^{2+}

As shown in Figure 5.36, the flux decline was almost identical in the two experiments. The flux decline from the experiment with settling was slightly less than that without settling, but the difference is infinitesimal. The results imply that a high concentration of particles alone is not responsible for membrane fouling. To foul the membrane, the particles must possess certain properties. For instance, the flocs in the experiment with the softened but not settled Missouri River water might be big enough to allow substantial back transport, and thus have little effect on the membrane fouling. Unfortunately, however, the comparison of these two experiments cannot lead to the desired clarity, because the DOC in raw (and softened) water from the two experiments was considerably different, with a higher DOC in the experiment with settling. Hence, the results for flux decline might represent combined effects of less particulate matter and more DOC in the settled water.

Efficiency of Cleaning Methods on the Missouri River Water

The efficiency of each cleaning procedure was analyzed for the raw Missouri River water and the softened water at both 90 and 165 mg/L CaO. The softened but not settled water at 90 mg/L CaO was also investigated. The results are shown in Figure 5.37. In general, surface wash was the most effective for all fouled membranes, especially for the membrane fouled with the raw Missouri River water. More than half of the lost flux was recovered by the surface wash in that membrane. The caustic wash showed no effects, or sometimes negative effects on recovering the lost flux. However, the acidic wash showed fairly good recoveries for all the membranes, which suggests that inorganic matter had some role in the fouling. The hardest fouling layer to recover by the cleaning methods was from the experiment with the softened but not settled water at 90 mg/L CaO. Therefore, it would be better to settle the softened water prior to ultrafiltration although the reduction of flux shows similar trends with or without settling.

Particle Size Distribution of the Missouri River Water

The particle number distribution of the raw Missouri River water was analyzed to examine particle deposition on the membrane surface. The result is shown in Figure 5.38. Most particles are in the range of 0.8 and 1.6 μm ($-0.08 < \log dp < 0.2$). Considering that the minimal back transport velocity was exhibited in the particles within the range of 0.3 to 1 μm diameter under the typical operational condition in this research, the particles in the raw Missouri River water showed a great tendency to be deposited on the membrane surface. Unlike kaolin (which shows no virtual fouling with the turbidity of 720 NTU), the raw Missouri River water showed a substantial flux decline. Therefore, changes from the particles with the minimal back transport velocity to the bigger flocs by softening might be one of the main reasons for the flux improvement at the lime dose of 90 mg/L CaO.

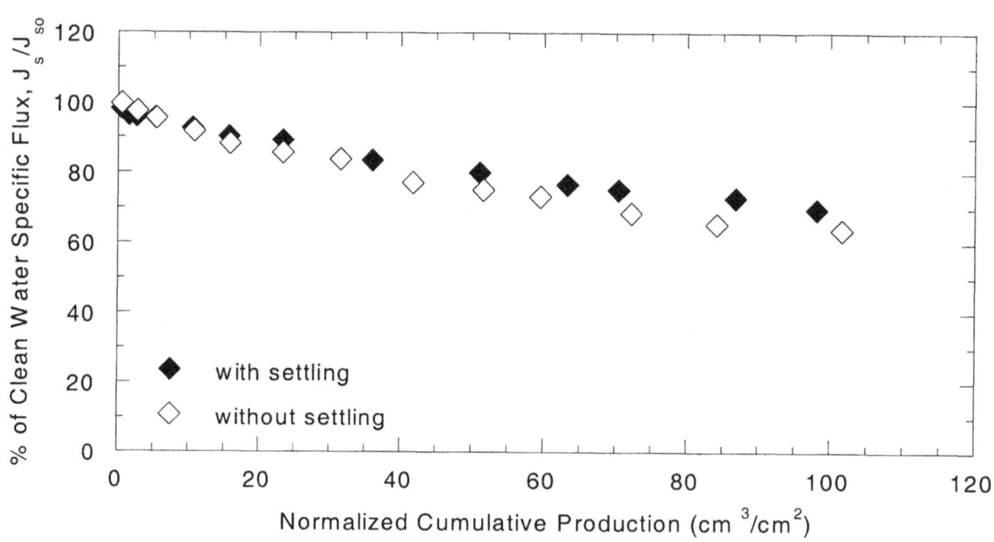

Figure 5.36 Flux decline: effects of settling (Missouri River water softened at 90 mg/L CaO)

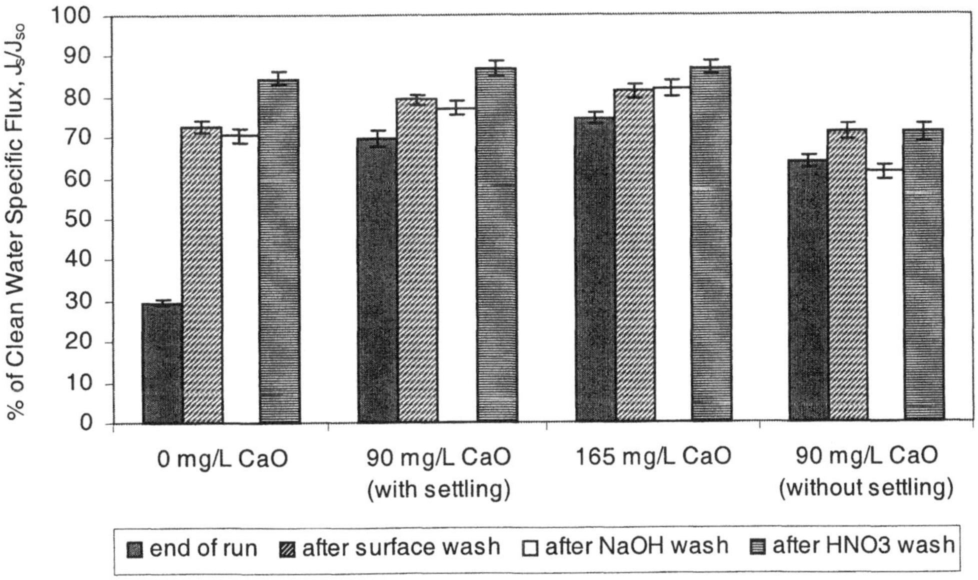

Figure 5.37 Membrane cleaning: Missouri River water at different conditions

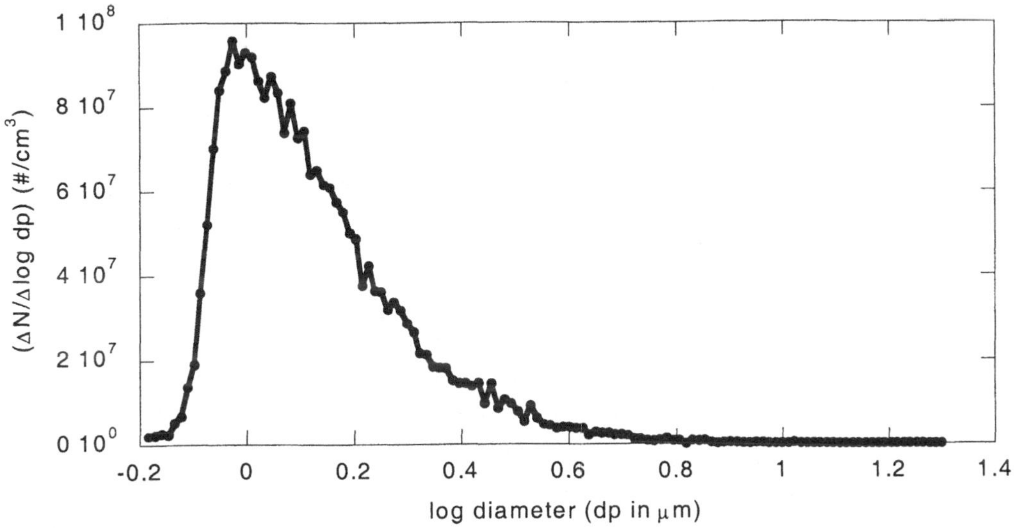

Figure 5.38 Particle number distribution of the raw Missouri River water

SEM of Membrane Surfaces Fouled by the Missouri River Water

The SEM images from the membrane fouled with the raw Missouri River water are presented in Figure 5.39. Many particles and organic matter cover the surface area. Particles have a variety of shapes including flat sheet-like, round, and blade-like shapes. Many particles were approximately 1 μm in size, which was consistent with the result from the particle size distribution by the Coulter Counter. Organic matter is shown as a lot of fibril shapes and is usually associated with the particles.

The images from the raw Missouri River water show a similarity to the images from the synthetic water with dextran and kaolin (presented in Figure 5.28). The numerous particles and fibril shapes of organic matter are common in the images from both membranes. However, the flux decline was tremendously different, which illustrated that the direct images of the fouled membrane by SEM are insufficient to understand fouling in the membrane processes.

The images of foulants after softening the Missouri River water at the lime dose of 90 mg/L CaO are shown in Figure 5.40. Similar to the images from the raw water, many solid deposits associated within a fibril network of organic matter are extensively shown in the SEM images. However, solids seem to have lumped together, and thus have less discrete edges compared to the particles in the raw water, which have more clear-cut edges. In addition, image (d) showed a place at which a lot of solids were stuck together and made a big hump on the membrane surface. If the NOM acted as glue within the hump, cleaning with the caustic wash solution might be unable to remove organic matter because the solids would coagulate more under the high pH and prevent the wash solution from reaching the organic matter.

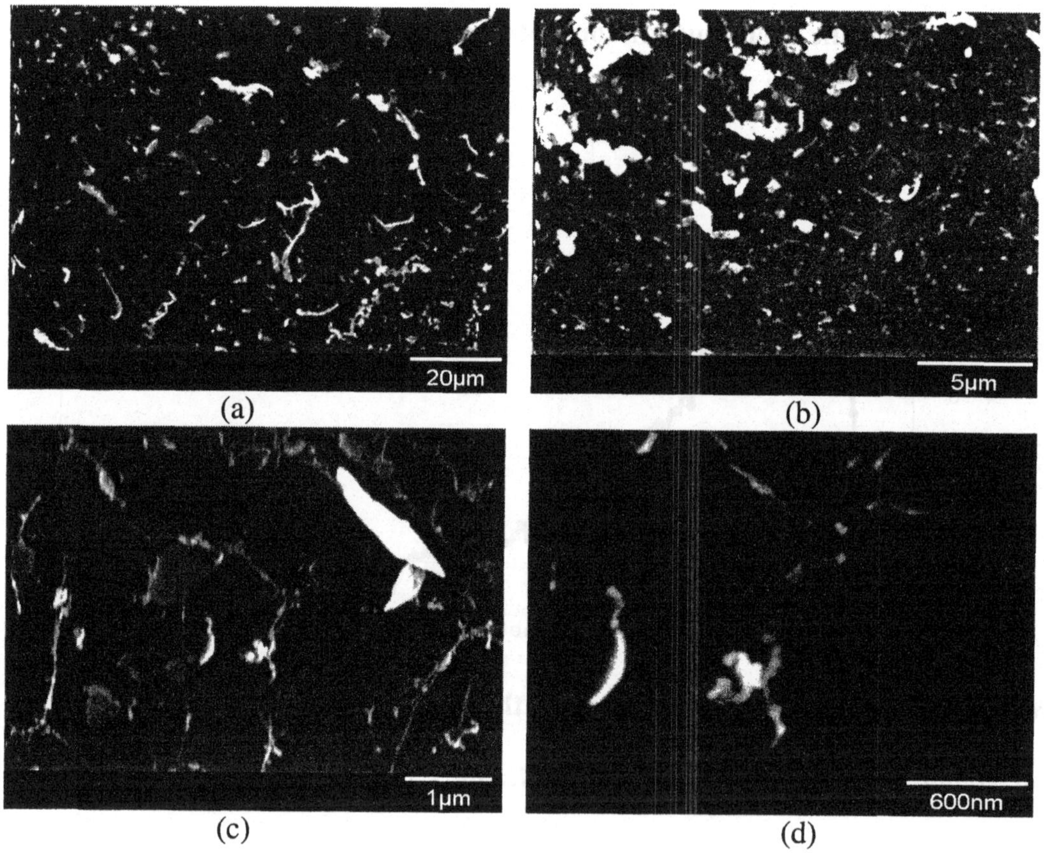

Figure 5.39 SEM images of the membrane fouled by raw Missouri River water ([a] 1000x, [b] 5000x, [c] 20000x, and [d] 50000x resolution)

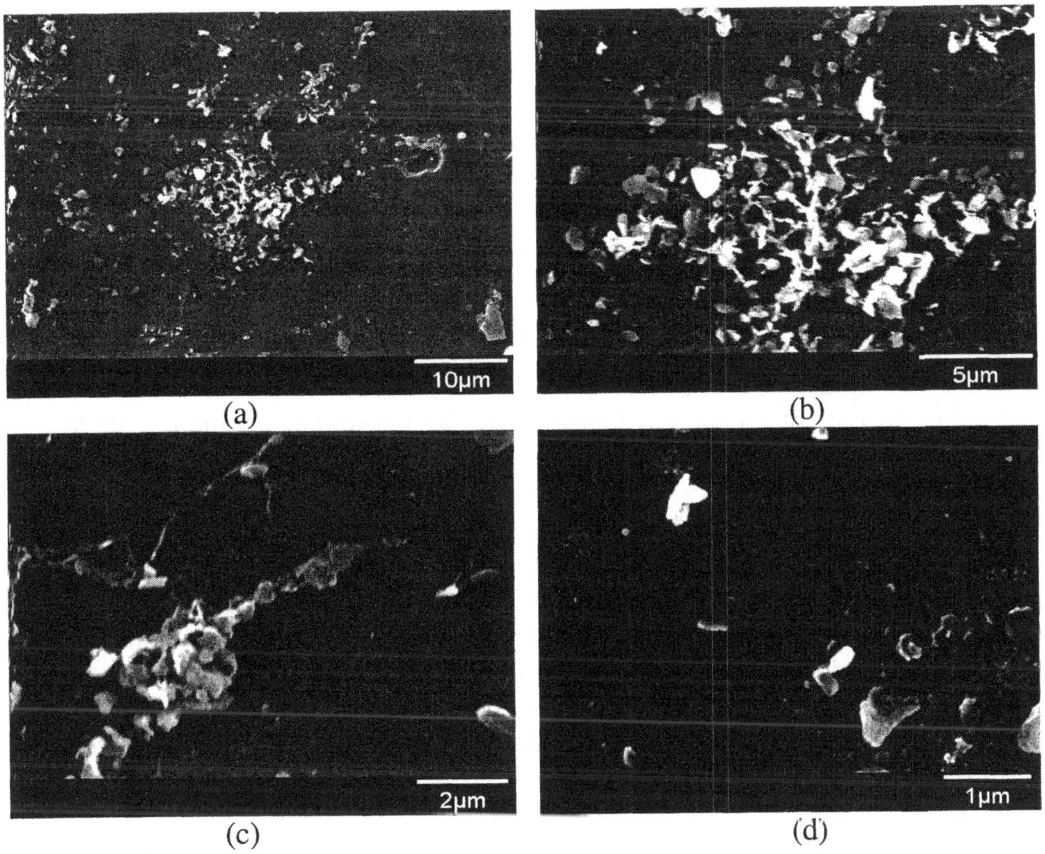

Figure 5.40 SEM images of the membrane fouled by the Missouri River water softened at 90 mg/L CaO and 3 mg/L Fe ([a] 2000x, [b] 5000x, [c] 10000x, and [d] 20000x resolution)

Table 5.31
Atomic composition: effects of softening Missouri River water

Lime dose (mg/L CaO)	C(1s) (%)	O(1s) (%)	S(2p) (%)	Ca(2p) (%)	Mg(2s) (%)
0	56.4	39.2	3.5	0.4	0.6
90	47.5	48.4	1.8	0.9	1.5

XPS of Membrane Surfaces Fouled by the Missouri River Water

The X-ray photoelectron spectroscopy (XPS) was also performed for the membrane fouled with the Missouri River water to measure the atomic composition of the top layer of the membrane. The concentrations of several elements based on XPS peak are presented in Table 5.31. Two membranes, from the experiments with the raw water and the water softened at the lime dose of 90 mg/L CaO, were analyzed. In general, the composition of O(1s) showed a higher fraction (i.e., 40 to 50%) in the membranes fouled with the Missouri River water than those fouled with other water sources such as Lake Austin water and the synthetic organic water with dextran (i.e., approximately 20 to 27%).

Detailed spectra of C(1s) and O(1s) from the membranes fouled with Missouri River water are presented in Figure 5.41. As shown in the composition results, the fouled membrane had much lower intensities of C(1s) but much higher intensities of O(1s) than that from the clean membrane. The spectrum of C(1s) from the membrane fouled with raw Missouri River water was quite similar to that fouled with raw alginic acid because of the significantly long tailed shape. Like the raw alginic acid, three binding energies (i.e., 285.1, 286.2, and 287.2 eV) were identified by the Gaussian curve fitting method. The presence of the third binding energy indicated that more oxidized C bonding was abundant in the raw Missouri River water, since the binding energy shifts to the higher energy with increasing oxidation state (Briggs 1983). After softening at the lime dose of 90 mg/L CaO, the spectrum of C(1s) from the membrane fouled with the softened Missouri River water became flat like a plateau, illustrating a similar intensity in the range between 284.6 eV and 287.4 eV of binding energy. The spectrum was quite different from that fouled with alginic acid, which showed that the intensity at the higher binding energy than 286.5 eV became close to that of the clean membrane with increasing degree of softening (Figure 5.21 (a)). The increased intensity of C(1s) at the binding energy of approximately 288.0 eV was consistent with the increased intensity of O(1s) at the binding energy of 533.2 eV, which is ascribed to C – O bond around the O atom. Similar to the extent of decreases in the C(1s), increases in the O(1s) were quite substantial in the Missouri River water as lime doses increased.

Considering the flux decline and XPS spectra from all the experiments, it seems that the flux decline was reduced when the intensity of C(1s) at the binding energy of 285.1 eV was decreased and the intensity of O(1s) at the binding energy of 532.0 eV was increased. This idea seems to hold true for many experiments that showed a dramatic reduction of flux decline, including the experiments with dextran softened at the lime dose of 170 mg/L CaO, with the Lake Austin water softened at the lime dose of 230 mg/L, and with the Missouri River water softened at the lime dose of 90 mg/L. All those experiments showed quite sudden reduction of flux decline and clearly demonstrated lower C(1s) and higher O(1s) intensities at the binding energy (*i.e.*, 285.1 eV and 532.0 eV) than other experiments. However, the intensities of C(1s) and O(1s) showed no particular pattern in the membranes fouled with alginic acid, which showed a gradual reduction of flux decline. These differences suggest that different fouling mechanisms occurred in these two cases.

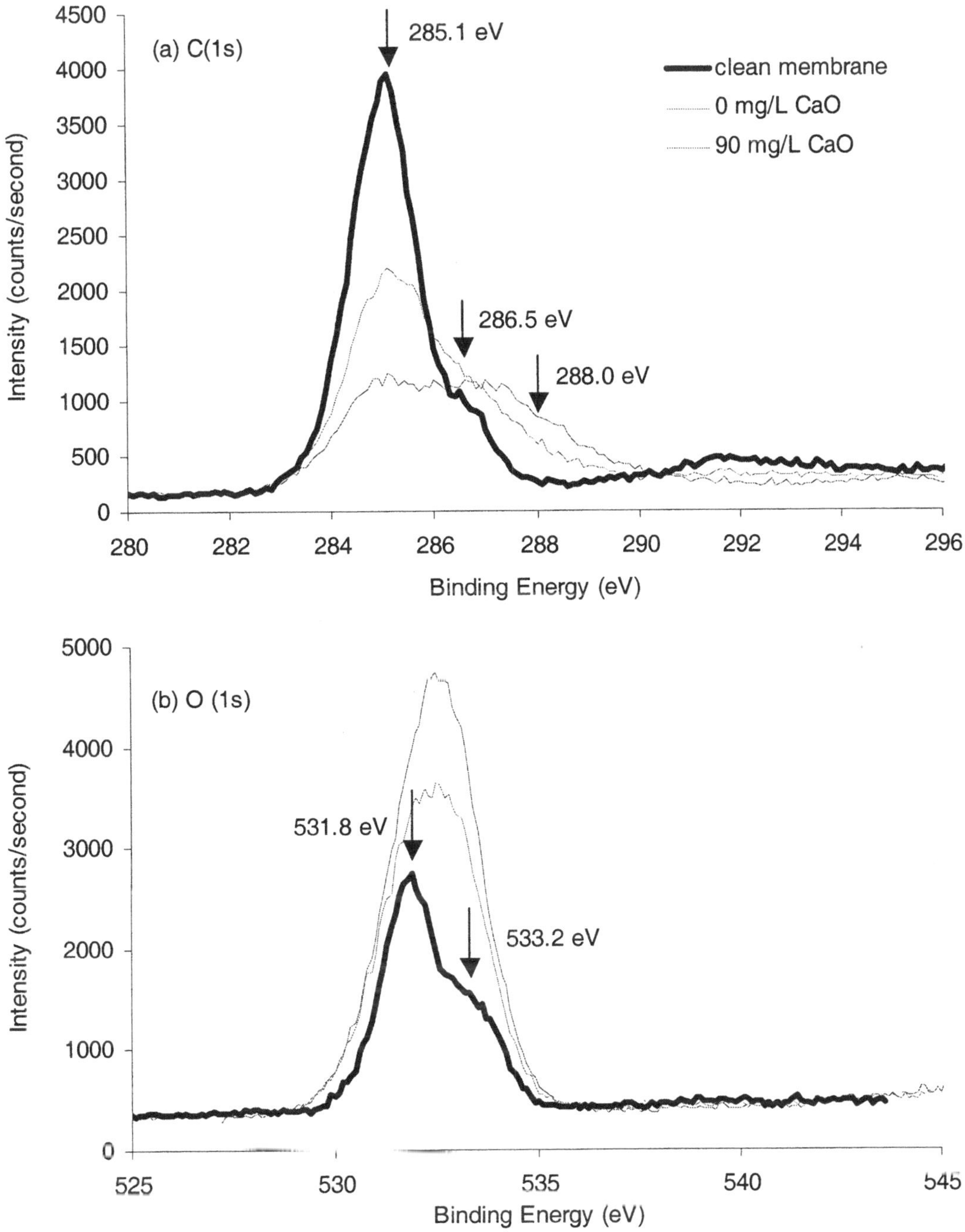

Figure 5.41 Spectra of the membrane fouled with Missouri River water at different lime doses

5.6 SUMMARY ON MEMBRANE FOULING

Two natural water sources were selected with different particle concentrations, Lake Austin water for a low turbidity water and the Missouri River water for a high turbidity water. Both water sources are generally hard waters with considerable concentrations of Ca^{2+} and Mg^{2+} and have similar DOC concentrations. Therefore, it might be said that Lake Austin water corresponds to the synthetic organic water with either dextran or alginic acid since little effect is expected from particle matter. The Missouri River water could correspond to the synthetic water with kaolin and dextran.

The flux decline in the experiments with Lake Austin water showed that the flux improvement was very limited to the enhanced softening condition and was dramatic at the Mg softening condition. In comparison, both synthetic organic waters (i.e., with either dextran or alginic acid) showed that the flux gradually increased with increasing lime doses and sometimes achieved approximately 100% of the clean water specific flux at the high lime dose. The molecular weight distributions of dextran solutions at the three lime doses (the same lime doses as Lake Austin water) clearly showed that the high molecular weight fraction of organic matter was preferentially removed in softening. However, the raw water and the softened Lake Austin at the lime dose of 170 mg/L showed that softening made a minor change in molecular weight distribution although it shifted a little to the lower molecular weight. The hydrophobic DOC analyses of the raw and the softened Lake Austin water revealed that the hydrophobic DOC was clearly correlated with the flux improvement. The hydrophobic fraction was slightly changed at the lime dose of 170 mg/L, and then drastically reduced at the highest lime dose (the same trend as the flux improvement at three lime doses).

The flux decline in the experiment with the raw Missouri River water obviously showed that the natural particles severely fouled the membrane since the big differences between Lake Austin water and the Missouri River water were in turbidity. However, the addition of kaolin to the synthetic organic water with dextran showed no differences from those without kaolin.

The flux improvement by enhanced softening was remarkable in the experiments with the Missouri River water. The improvement was not from the removal of high turbidity from the settling process because the flux in the experiment with the softened but not settled Missouri River water at 90 mg/L CaO was almost the same as that with settled water. Therefore, the flux improvement seems to stem from combination of two effects: the shift of the particle size distribution away from the region with the minimal back transport velocity, and the preferential removal of the hydrophobic DOC, which plays a big role in membrane fouling.

The synthetic water made with kaolin and dextran was designed to be similar to the Missouri River water. Without softening, the natural Missouri River water gave a much greater flux decline than the synthetic water. For both the Missouri River water and the synthetic water, softening (with or without settling) substantially reduced the flux decline. Apparently, the increased particle size and reduced organic concentration after softening was responsible for the flux improvement. But the greater flux decline of the natural water in comparison to the synthetic water was not changed by softening; the combination of smaller particles and greater hydrophobicity of the natural water was the cause of the greater fouling.

CHAPTER 6
CONCLUSIONS

Integrated water treatment with softening and ultrafiltration is proposed as a promising option for hard waters as a means to meet the changes in the recent drinking water regulations: the Disinfection/Disinfectant By-Product Rule and the Interim Enhanced Surface Water Treatment Rule. Although ultrafiltration can accomplish excellent disinfection by removal of microorganisms and some removal of natural organic matter (NOM) to meet the regulations prior to disinfection, fouling is a major potential impediment to its application. To investigate the feasibility of the softening process as a pretreatment for ultrafiltration, this research was designed to understand the nature of the fouling mechanisms for ultrafiltration membranes when used for waters that either require softening or have been softened in a precipitation process.

Four causes of fouling were hypothesized to be possible in the integrated water treatment with softening and ultrafiltration: inorganics that could form precipitates, organics, particles, and the combined fouling by particulate and organic matter. Two natural waters, Lake Austin water and Missouri River water, were characterized in terms of inorganic (especially hardness ions and alkalinity) and organic constituents. Based on the results from the characterization of the natural waters, synthetic waters were made to simulate those natural waters; inorganic and organic matter concentrations represented Lake Austin water and, when particles were involved, the turbidity concentration modeled Missouri River water. To investigate systematically the four causes of fouling, several synthetic waters were used: synthetic water only with inorganic constituents (synthetic inorganic water) for inorganic fouling, water with polysaccharides in the synthetic inorganic water for organic fouling, water with kaolin in the synthetic inorganic water for particle fouling, and water with kaolin and polysaccharides for the combined fouling. Using the two natural waters, three extents of softening were determined: standard softening, enhanced softening, and Mg softening conditions. In addition, scenarios of softening were investigated with synthetic inorganic water and Lake Austin water to determine a specific process with the least fouling. The three degrees of softening and the specific softening process were continuously used as extents of pretreatment prior to ultrafiltration. Fouling in ultrafiltration was examined with the reduction of flux and efficiency of recovery by three different cleaning methods. In addition, scanning electron microscopy (SEM) and X-ray photoelectron spectroscopy (XPS) were used to aid understanding of fouling that occurred on the membrane surface.

Conclusions based on the thorough investigations of the four fouling mechanisms follow.

1. Inorganic fouling shows negligible effect on membrane fouling. Inorganic fouling by $CaCO_3$ and $Mg(OH)_2$ precipitates was investigated in three aspects: effects of the slower precipitation kinetics in softening, the degree of softening, and various scenarios of softening such as lime softening alone, lime/soda ash softening, and lime/soda ash softening with pH adjustment. In all of these cases, ultrafiltration showed little flux decline. The experiment to investigate the effects of the slower precipitation kinetics showed that precipitation could be initiated in the membrane system. However, the precipitation occurred on the surface, not inside the pores, and the precipitates were big enough not to cause a resistance to water flux in ultrafiltration.

2. Ultrafiltration is not susceptible to much particulate fouling under the operational conditions in this research. Ultrafiltration showed virtually no flux decline when the synthetic inorganic water with only a clay material (without NOM) was treated. It should be noted that ultrafiltration sustained almost 100% of the clean water specific flux even when a highly turbid water with particles possessing the minimum back transport velocity was applied.
3. Fouling by both inorganic precipitates and particulate matter (a clay material) shows a negligible effect on reduction of flux in membranes. A substantial amount of solid deposits were observed by the SEM images from fouled membranes. However, the results of the SEM combined with the flux experiments clearly showed that a layer with only inorganic particles created by either precipitates from softening or a clay material (kaolin without NOM) exhibited no resistance to the water flow in the membrane surfaces.
4. Organic matter, either natural organic matter or a simple organic component used as a NOM surrogate, is the most detrimental foulant on the membrane process. Two polysaccharides, dextran and alginic acid, were selected to investigate organic fouling, since both compounds showed similar trends of inorganic and DOC removals as Lake Austin water. The flux decline was significant when raw waters (without softening) were applied regardless of the water source, i.e., natural or synthetic organic waters with dextran and alginic acid. The experiments with polysaccharides showed that fouling occurred severely when organic matter had a relatively high molecular weight compared to the molecular weight cut-off of the membrane.
5. Softening pretreatment significantly reduces organic fouling in ultrafiltration. For each water, increasing the extent of softening continually reduced the fouling, but the extent of the improvement of water flux varied depending on the feed water. For instance, significant flux improvement was achieved at the Mg softening condition for Lake Austin water but the standard and enhanced softening condition was sufficient for Missouri River water. The extent of flux improvement with increasing the extent of softening is summarized in Table 6.1. The flux was reported at the normalized cumulative production of 60 cm^3/cm^2 for comparison. As the degree of softening increased, a dramatic improvement in the flux reduction occurred with dextran, Lake Austin water, and Missouri River water while a more gradual flux improvement was observed with alginic acid.
6. The cause of flux improvement in the experiments with the polysaccharides (dextran and alginic acid) likely stemmed from the preferential removal of its high molecular weight fraction in softening. However, Lake Austin water consisted of relatively low molecular weight organic matter; therefore, softening induced small changes in its molecular weight size distribution. Rather, changes in hydrophobicity with softening were closely associated with the flux improvements of Lake Austin water. Therefore, softening has an ability to reduce foulants prior to ultrafiltration by preferential removal of both the high molecular weight fraction and hydrophobic portions of NOM. The extent of removal of these constituents (which could overlap to a large extent for some waters) is dependent on the water characteristics.

Table 6.1
Percent of clean water specific flux at the normalized cumulative production of 60 cm^3/cm^2

<table>
<tr><th rowspan="2">Water sources</th><th colspan="4">% of clean water specific flux</th></tr>
<tr><th>Raw</th><th>Standard Softening</th><th>Enhanced Softening</th><th>Mg Softening</th></tr>
<tr><td>Dextran</td><td>60</td><td>75</td><td>90</td><td>90</td></tr>
<tr><td>Alginic acid</td><td>20</td><td>45</td><td>55</td><td>65</td></tr>
<tr><td>Lake Austin</td><td>70</td><td>72</td><td>75</td><td>95</td></tr>
<tr><td>Missouri River</td><td>40</td><td colspan="2">75</td><td>80</td></tr>
</table>

7. The effects of particle fouling on membrane were shown in the results from experiments with synthetic water mixed with dextran and kaolin as well as with Missouri River water. Kaolin was added to investigate the combined fouling by particulate and organic matter (dextran). The combined fouling showed no difference from the fouling with only organic matter. The similarity might stem from the fact that polysaccharide has no ability to be adsorbed on the clay material, so that the polysaccharide behaves the same with respect to the membrane with or without clay material being present. The effects of fouling by natural particles were shown with experiments with Missouri River water. The much greater flux decline with Missouri River water than Lake Austin water indicates the importance of particle fouling since the big difference between Lake Austin water and Missouri River water is the turbidity. While the synthetic particles alone showed a negligible effect, the natural particles showed a dramatic effect on membrane fouling. The interpretation implies that other organic components, which can be adsorbed on the clay material, would act differently from the polysaccharide.
8. Softening also reduces fouling and improves the water flux in ultrafiltration membranes when a highly turbid water such as Missouri River water is treated. The improvement of flux was significant even with the standard and enhanced softening condition for Missouri River water. The improved flux at this condition implies that softening changes the particle and NOM properties; particles are converted to bigger flocs by coagulation, particle composition is changed from natural particles to the $CaCO_3$ and $Mg(OH)_2$, and NOM becomes associated with the fresh precipitates. After softening, particles apparently do not cause much membrane fouling; even without solid-liquid sedimentation, very little flux decline occurred after softening at the standard condition. The results also suggest that the DOC fraction that is preferentially removed in softening has a greater role in fouling of ultrafiltration membranes than the amount of DOC remaining after the softening process.
9. Three cleaning methods applied at the end of each run effectively reveal different fouling mechanisms. Dextran was likely to chemically attach to the membrane since the surface wash showed little effect and the caustic wash showed a relatively significant effect on the recovery, whereas substantial amounts of alginic acid seemed to be deposited on the membrane surface since the surface wash recovered most of the lost flux. In addition, the relatively successful recovery by the acidic wash implies that alginic acid is bound with inorganic matter to some extent. Three cleaning methods were less effective in recovering the lost flux from the membranes fouled

with the two natural waters than those fouled with synthetic organic water, dextran and alginic acid.

10. The SEM images of the surface of the membranes are quite valuable in understanding the identities of foulants. However, the SEM images sometimes fail to predict the reduction of flux; therefore, other measurements are necessary to investigate actual fouling phenomena in the membrane processes. The XPS was beneficial in determining the relative composition of each fouling material on the membrane surfaces.

Based on the above discussions, the integrated treatment with softening and ultrafiltration proves to be a promising option for hard waters. The degree of softening required to improve water flux should be determined with the raw water to be applied because it depends on the raw water characteristics.

6.1. RECOMMENDATIONS

Some recommendations for future work to better understand the fouling mechanisms in integrated treatment with softening and ultrafiltration stem from this research. Other compounds beside polysaccharides could be investigated. The polysaccharides are hydrophilic and show no adsorption to the clay materials. Therefore, other organics with more hydrophobic properties might reveal different fouling phenomena in the integrated treatment. In addition, performing separate experiments with softening different NOM fractions, such as hydrophobic and hydrophilic fractions, might be useful to better understand organic removal in softening and fouling in the membrane processes. More research should be performed to understand removal of foulants in the softening process because removal efficiency in the pretreatment process controls performance in the subsequent ultrafiltration.

The primary membrane material in this research was polysulfone. Regenerated cellulose was also investigated at the beginning of experiments. Since this material is quite hydrophilic, little flux decline was observed regardless of different pretreatment conditions and more experiments were not performed. Other membrane materials might show different fouling phenomena. Therefore, research with other membrane materials is required to better understand fouling in the integrated treatment process with softening and ultrafiltration.

Finally, in this research, the operational period is relatively short compared to the operation in real systems. Therefore, research using long-term operation might be necessary before results of this research can be applied to the real world.

APPENDIX A
LIQUID SAMPLE ANALYSIS

This section was adapted from the section "Materials and Methods in Ralls (1999), with permission.

A.1 PH

pH measurements were taken using an Orion Research Model 701A pH Meter with an Orion Sureflow Ross Combination pH Probe. Reported results are accurate to within ±0.01 pH unit. Before taking measurements, the meter was calibrated using buffer solutions. For measuring the pH of softened samples, the meter was calibrated using pH 7 and pH 10 buffer solutions. For measuring the alkalinity of softened samples, the meter was calibrated using pH 4 and pH 7 buffer solutions.

A.2 ALKALINITY

Alkalinity is a measure of the ability of a water to buffer against changes in pH. The solubilities of Ca^{+2} and Mg^{+2} (and therefore their removals by precipitation) are highly pH-dependent, so alkalinity is an important measure for characterizing a water before and after softening. Alkalinity was measured using the titrimetric approach in accordance with Standard Method 2320B (APHA et al. 1998), except that a 50-mL sample size was used. The 0.01 N sulfuric acid (H_2SO_4) titrant was prepared by diluting 50 mL 0.1 N H_2SO_4 to 500 mL with distilled deionized water. The titrant strength was periodically checked by standardization against 40.0 mL of 0.05 N sodium carbonate solution in accordance with Standard Method 2320B.

A.3 CALCIUM AND MAGNESIUM

Ca^{+2} and Mg^{+2} are the primary hardness ions, and the goal of conventional softening is to remove them through precipitation, making the measurement of Ca^{+2} and Mg^{+2} important for characterizing a water before and after softening. Ca^{+2} and Mg^{+2} were measured using a Perkin Elmer 2380 Flame Atomic Absorption Spectrophotometer (FAA), in accordance with Standard Method 3111 (APHA et al. 1998).

Ca+2 Water Standards

Prior to analyzing Ca^{+2} samples, the FAA was calibrated by analyzing water standards at concentrations of 0, 3, 6, 9, 12, and 15 mg/L Ca^{+2}. The FAA response to water standards containing known concentrations of Ca^{+2} was used to determine the concentrations present in the softened water samples. The water standards were created from a 500 mg/L Ca^{+2} stock standard (SS) in three steps. First, a 100 mg/L Ca^{+2} primary dilution standard was created by diluting 20 mL of the SS to 100 mL with distilled water. Second, the five standards were created by diluting the primary dilution solution to 100 mL with distilled water. The volumes used to create the standards are listed in Table A.1. Finally, 10 mL lanthanum oxide solution (see below) was added to each standard to prevent phosphate interference. The standards were stored in 125-mL plastic bottles with screw caps.

Lanthanum Oxide Solution

A solution of lanthanum oxide (La_2O_3) was added to all calcium standards and samples to prevent phosphate interference. The solution was created in two steps. First, 29.33 g La_2O_3 was dissolved in 125 mL concentrated hydrochloric acid (HCl). Second, approximately 300 mL distilled water were added to a 500 mL volumetric flask and the flask was placed in a cooling bath. Taking care to prevent the flask from overheating, the La_2O_3/HCl solution was slowly added. The flask was then filled to volume with distilled water.

Mg+2 Water Standards

Prior to analyzing Mg^{+2} samples, the FAA was calibrated by analyzing water standards at concentrations of 0, 0.3, 0.6, 0.9, 1.2, and 1.5 mg/L Mg^{+2}. The FAA response to water standards containing known concentrations of Mg^{+2} was used to determine the concentrations present in the softened water samples. The water standards were created in three steps. First, a 100 mg/L Mg^{+2} stock solution was created by dissolving 0.1658 g magnesium oxide (MgO) in a minimum amount of 1+1 nitric acid (see below), adding 10 mL concentrated nitric acid (HNO_3), and diluting to 1 L with distilled water. This dilution was performed by adding approximately 900 mL distilled water to a 1-L volumetric flask and placing the flask in a cooling bath. Taking care to prevent the flask from overheating, the MgO/HNO_3 solution was slowly added. The flask was then filled to volume with distilled water.

The stock solution was then used to create a 10 mg/L Mg^{+2} primary dilution solution by diluting 10 mL of the stock solution to 100 mL with distilled water. Finally, the water standards were created by diluting the primary dilution solution to 100 mL with distilled deionized water. The volumes used to create the standards are listed in Table A.2. The standards were stored in 125-mL plastic bottles with screw caps.

Table A.1
Ca^{+2} Water Standards

Standard Concentration (mg/L Ca^{+2})	Primary Dilution Solution Volume* (mL)
0	0
3	3
6	6
9	9
12	12
15	15

Table A.2
Mg+2 Water Standards

Standard Concentration (mg/L Mg^{+2})	Primary Dilution Solution Volume* (mL)
0.0	0
0.3	3
0.6	6
0.9	9
1.2	12
1.5	15

1+1 Nitric Acid

A solution of 1+1 nitric acid was prepared by adding 5 mL concentrated HNO_3 to 5 mL distilled water.

Sample Preparation

Dilution factors were selected based on previous experience and samples were diluted to 100 mL with distilled water and stored in 125-mL plastic bottles with screw caps. Ten milliliters La_2O_3 solution were added to Ca^{+2} samples after dilution to prevent phosphate interference.

Analysis

Flame atomic absorption operates on the principle that metals in a reducing environment will go to their elemental state, and their outer electrons will be excited to higher energy levels by gaseous collisions. As the electrons fall back to ground state, radiation is emitted. However, only about 1% of the atoms are in an excited state; the rest are in ground state, where they absorb radiation. This absorbance is measured by an atomic absorption spectrophotometer.

Specifically, an aqueous sample is sucked into a flame by an aspirator, which consists of gas flowing past the sample and drawing a fraction of the liquid into the flame as small droplets. The fuel-rich primary combustion zone of the flame turns atoms of easily reduced elements in the aerosol sample into gaseous atoms in their reduced state. Radiation is directed through the primary combustion zone flame by a hollow cathode lamp. The cathode of the lamp is constructed of (or plated with) the metal being measured. When current is applied across the lamp electrodes, the metal atoms are excited and emit their characteristic radiation as they fall back to ground state. This radiation, directed through the flame, is absorbed by the metal atoms in the sample and re-emitted in any direction. A photomultiplier tube measures the intensity of radiation reaching it. Every element exhibits maximum absorption at specific wavelengths, which is dialed into the FAA during set up. The maximum absorbances for Ca^{+2} and Mg^{+2} are found at wavelengths of 422.7 nm and 285.2 nm, respectively.

Some radiation is emitted by the flame, so to distinguish between the radiation from the flame and the radiation from the lamp, a fan is located in front of the lamp, causing the lamp's signal to fluctuate. The spectrophotometer can then determine the magnitude of the signal from the flame and subtract it from the combined signal to determine the signal from the lamp and sample alone. To determine the signal from the sample alone, the 0 mg/L sample is measured and zero absorbance was set, effectively subtracting out the signal from the lamp.

It should be noted that this is a total element analysis; that is, the element may be in several different oxidation states in the water, but it is reduced to ground state in the gas phase, so that all species are measured.

A.4 ULTRA-VIOLET LIGHT ABSORBANCE

UV absorbance at a wavelength of 254 nm is often used to approximate the concentration of dissolved organic carbon in a sample. In this research, UV absorbance was measured using a Perkin-Elmer Lambda 38 UV/Visible Spectrophotometer with a 5-cm cuvette or Agilent 8453E UV-visible Spectroscopy with 1-cm cuvette. Quality assurance tests were performed on several waters to ensure that measurements from the two different analyzers were consistent, as shown in Table A.3.

The spectrophotometer operates by transmitting light at a specific wavelength (entered via a keypad mounted on the unit) through a sample. On the other side of the sample cell from the lamp, a photomultiplier tube compares the intensity of light detected through the sample to the intensity transmitted through a blank (a cuvette filled with distilled deionized water). Prior to analyzing a sample, the spectrophotometer was calibrated to zero using distilled deionized water in the sample cuvette.

A.5 ORGANIC CARBON

Organic carbon is frequently used as a measure of the concentration of NOM present in a water. Three different organic carbon measurements are total organic carbon (TOC), dissolved organic carbon (DOC), and hydrophobic DOC. The analytical procedures for all three are identical except for sample preparation. No special preparation is necessary for TOC, which is a measure of both dissolved and particulate organic carbon. To determine DOC, the sample is filtered prior to measuring organic carbon. To measure hydrophobic DOC, part of the filtered sample is analyzed for DOC and part is passed through a column packed with ion-exchange resin to remove the hydrophobic fraction of the organic carbon by sorption onto the resin (as explained in the Section 3.5.2).

Table A.3
Comparison of UV absorbance from two analyzers

Sample	UV absorbance (1/cm)	
	Perkin-Elmer Lambda 38	Agilent 8453E
#1	0.072	0.0724
#2	0.060	0.0615
#3	0.060	0.0618

Glassware Preparation

Glassware for organic carbon analysis was prepared by washing with soap and water, rinsing three times with tap water and three times with distilled water, and soaking in a bath of dilute nitric acid (see below) for two hours. Following the acid bath, glassware was rinsed with distilled/deionized water three times. The sample vials were baked at 550°C for two hours, and brown bottles (used for reagents and standard solutions) were baked at 400°F overnight. Screw caps and Teflon-lined septa were washed with soap and water, rinsed with distilled water, rinsed with reagent-grade acetone and allowed to air-dry.

Dilute Nitric Acid

A nitric acid solution was prepared by adding approximately 800 mL distilled deionized water to a 1-L volumetric flask and placing the flask in a cooling bath. Taking care to prevent the flask from overheating, 100 mL concentrated HNO_3 were slowly added. The flask was then filled to volume with distilled/deionized water.

Water Standards

Water standards at concentrations of 1, 2, 3, 4, and 5 mg/L C were run prior to every analysis. They were created in three steps. First, a 1,000 mg/L C stock solution was created by dissolving 425 mg potassium hydrogen phthalate ($KHC_8H_4O_4$, previously dried at 103°C for two hours) in 200 mL of distilled/deionized water. The stock solution was then acidified with 100 μL concentrated phosphoric acid (H_3PO_4) and stored in a brown glass bottle with a Teflon-lined screw cap at 4°C for up to one month. Second, a 100 mg/L primary dilution solution was created by diluting 10 mL of the stock solution to 100 mL using distilled deionized water. Finally, the water standards were created by diluting the primary dilution solution to 500 mL with distilled deionized water. The volumes used to create the standards are listed in Table A.4.

For this research, a 5-point calibration was performed. The instrument response for each standard was recorded and plotted against its actual concentration. This plot was fitted with a regression line and the equation for this line was used during sample analysis to accurately determine sample concentrations.

Table A.4
Organic Carbon Water Standards

Standard Concentration (mg/L C)	Primary Dilution Solution Volume* (mL)
1	5
2	10
3	15
4	20
5	25

Sample Preparation

Prior to DOC analysis, a 50-mL glass vial was rinsed with sample and then filled with sample. Two drops of concentrated H_3PO_4 were added to prevent bacterial degradation of the sample and the vial was capped with a screw cap and Teflon-lined septum. Samples were stored at 4°C for no longer than 2 weeks before analysis.

Analysis

Organic carbon was measured first using a Dohrmann DC-180 TOC Analyzer, using the non-purgeable organic carbon (NPOC) setting and changed to a Tekmar-Dohrman Apollo 9000 TOC Combustion Analyzer. The instruments use basically the same method for sparging inorganic carbon and purgeable organic carbon and detecting the amount of CO_2 in samples. In both instruments, samples are acidified with 20% phosphoric acid (see below) and sent to a sparger, where the inorganic carbon and purgeable organic carbon are stripped by oxygen. A measured amount of sample is then delivered from the sparger to the reactor in the DC-180 TOC Analyzer, along with oxygen (supplied by a tank piped up to the analyzer) and 2% potassium persulfate (see below). In the reactor, the NPOC is oxidized to CO_2 by ultra-violet light and oxygen-rich persulfate. In the Appollo 9000 TOC Combustion Analyzer, samples are introduced to the combustor with temperature 800°C and the NPOC is oxidized to CO_2 in the presence of catalysts. An infrared detector produces an electrical signal that is proportional to the amount of CO_2 in the sample. This signal is integrated, scaled, and displayed as mg/L carbon (C). To ensure quality assurance, several samples were measured with both instruments. As shown in Table A.5, differences between two analyzer were quite small, i.e., within 10%.

Table A.5
Comparison of organic carbon concentration from two analyzers

Samples	DC-180	Apollo 9000	% difference
#1	10.89	10.30	5.56
#2	4.23	3.90	8.15
#3	3.93	3.86	1.86
#4	4.06	3.96	2.52
#5	3.36	3.12	7.60

20% Phosphoric Acid:

A solution of 20% phosphoric acid, used to acidify samples prior to sparging, was prepared by adding approximately 100 mL distilled deionized water to a 200-mL volumetric flask and placing the flask in a cooling bath. Taking care to prevent the flask from overheating,

40 mL of concentrated H_3PO_4 were slowly added. The flask was then filled to volume with distilled deionized water.

2% potassium persulfate:

A solution of 2% potassium persulfate, used to oxidize the NPOC to CO_2 in the reactor, was prepared by dissolving 20 g of potassium persulfate ($K_2S_2O_8$) in 1 L distilled deionized water, then adding 2 mL concentrated H_3PO_4.

In this research, the sample line was rinsed with sample prior to analysis, and each sample was measured five times. The first measurement usually diverged from the others and was always discarded. The four remaining values were generally in very good agreement; standard deviations were typically on the order of 0.05 mg/L C. DOC concentrations measured in this research were in the range of 2 to 5 mg/L C, so this standard deviation represents 1 to 2.5% of the sample concentration. The DOC concentration in the sample was determined based on the average value measured and the standard curve equation determined during calibration.

APPENDIX B
QUALITY ASSURANCE OF MEMBRANE EXPERIMENTS

B.1 SYNTHETIC INORGANIC WATER

Table B.1
Operational conditions of duplicated experiments with synthetic inorganic water (lime dose: 230 mg/L CaO)

Experiments	TMP (kPa)	Crossflow velocity (cm/s)	Clean water specific flux, J_{so} (L/m^2-hr-kPa)
#1	94.5 – 96.6	10.4	1.684
#2	89.0 – 101.4	10.3	1.900

Table B.2
Water quality of duplicated experiments with synthetic inorganic water (lime dose: 230 mg/L CaO)

Experiments	Sample	pH	Turbidity (NTU)	Ca^{2+} (mg/L)	Mg^{2+} (mg/L)
#1	Raw	8.54	-	57.6	19.1
	Softened	11.81	171.1	77.8	9.3
	Feed	11.67	5.05	43.5	3.2
	Filtrate	11.64	0.055	35.6	0.3
#2	Raw	8.31	-	36.0	16.5
	Softened	12.02	88.50	33.6	9.5
	Feed	-	4.15	31.9	0.4
	Filtrate	-	0.024	35.7	0.3

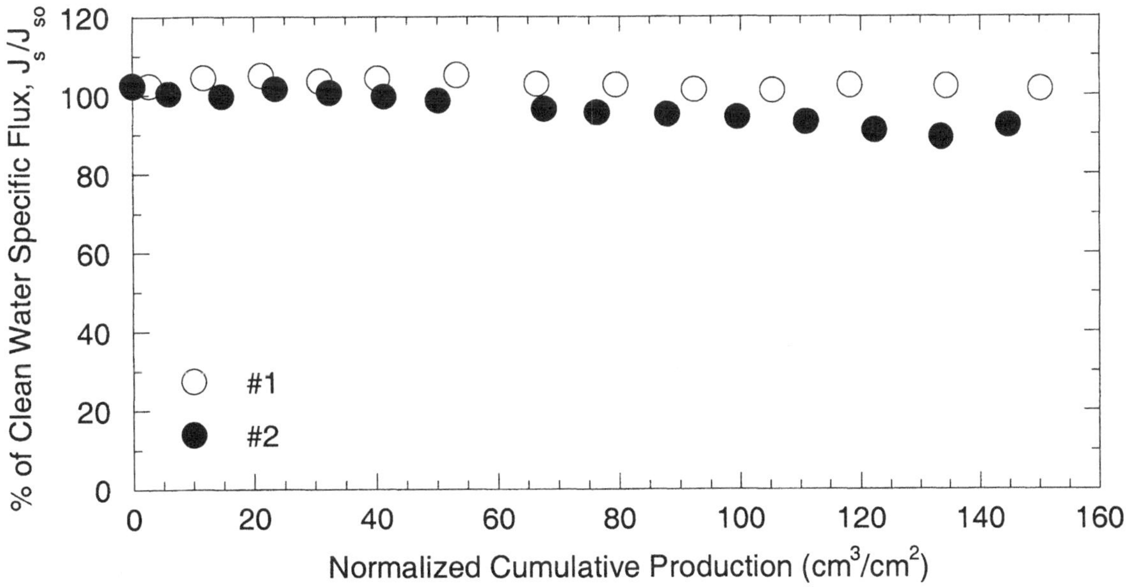

Figure B.1 Flux decline: duplicated experiments with synthetic inorganic water (lime dose: 230 mg/L CaO)

B.2 SYNTHETIC ORGANIC WATER WITH DEXTRAN

Raw Dextran

Table B.3

Operational conditions of duplicated experiments with raw dextran (without softening)

(Nominal molecular weight: 60 kDa, DOC: 4 mg/L)

Experiment	TMP (kPa)	Crossflow velocity (cm/s)	Clean water specific flux, J_{so} (L/m^2-hr-kPa)
#1	90.3 95.2	10.1	1.95
#2	89.7 – 93.1	10.1	1.94

Table B.4

Water quality of duplicated experiments with raw dextran (without softening)

(Nominal molecular weight: 60 kDa, DOC: 4 mg/L)

Experiment	Sample	pH	Ca^{2+} (mg/L)	Mg^{2+} (mg/L)	DOC (mg/L)
#1	Feed	8.26	48.2	16.9	5.4
	Filtrate	8.25	45.0	15.6	2.7
#2	Feed	8.30	50.0	17.4	4.8
	Filtrate	8.34	50.8	16.9	2.5

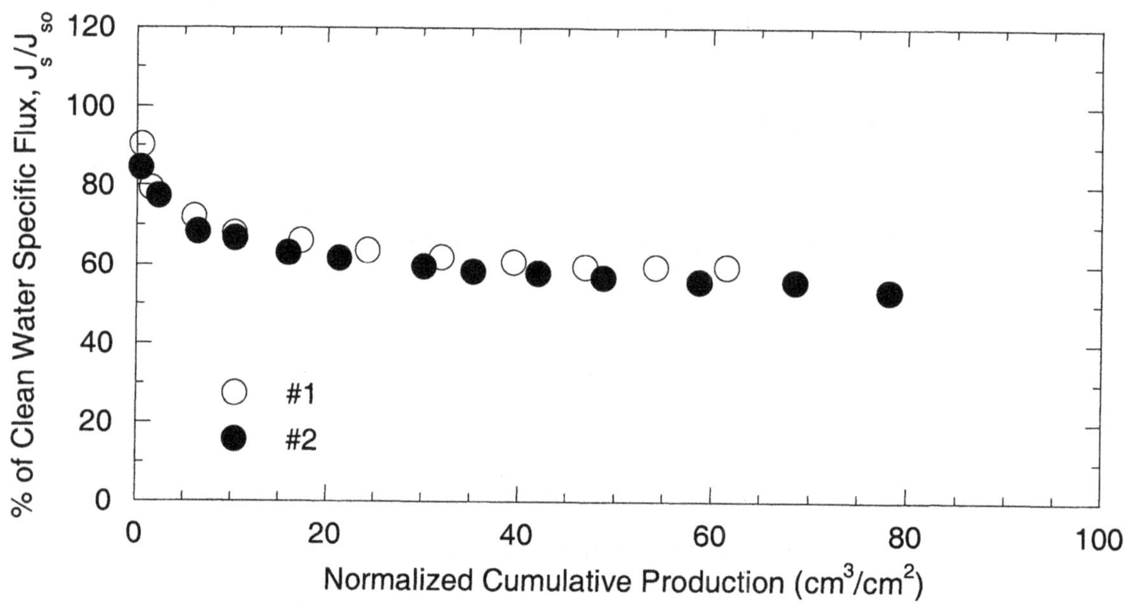

Figure B.2 Flux decline: duplicated experiments with raw dextran (without softening)
(Nominal molecular weight: 60 kDa, DOC: 4 mg/L)

Softened Dextran at 125 mg/L CaO

Table B.5

Operational conditions of duplicated experiments with softened Dextran60 at 125 mg/L CaO (4 mg/L C)

Experiments	TMP (kPa)	Crossflow velocity (cm/s)	Clean water specific flux, J_{so} (L/m^2-hr-kPa)
#1	89.7 – 93.1	10.0	2.16
#2	88.3 – 91.7	9.8	1.42

Table B.6

Water quality of duplicated experiments with softened Dextran60 at 125 mg/L CaO (4 mg/L C)

Experiments	Samples	pH	Turbidity (NTU)	Ca^{2+} (mg/L)	Mg^{2+} (mg/L)	DOC (mg/L)
#1	Raw	8.29	NA	50.8	17.0	5.8
	Softened	10.83	47.1	19.6	13.9	4.5
	Feed	10.56	2.03	12.8	13.6	4.1
	Filtrate	10.52	0.014	14.0	13.6	2.7
#2	Raw	8.45	NA	57.2	18.5	3.9
	Softened	10.68	33.6	14.0	15.8	2.8
	Feed	-	1.87	9.0	15.6	2.7
	Filtrate	-	0.008	12.8	16.1	1.6

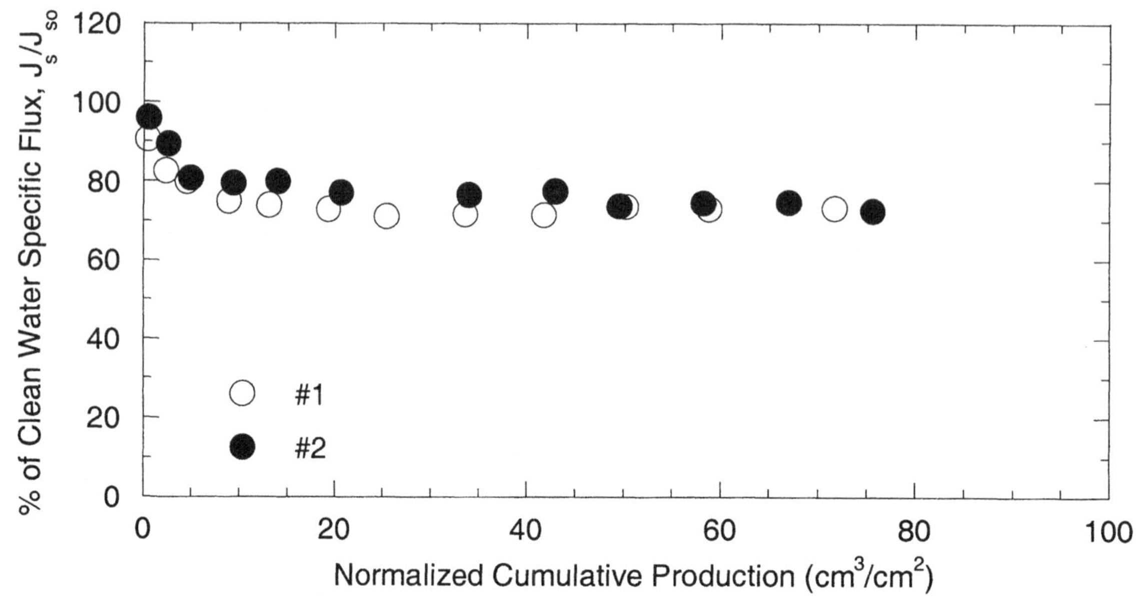

Figure B.3 Flux decline: duplicated experiments with softened Dextran60 at 125 mg/L CaO (4 mg/L C)

B.3 SYNTHETIC ORGANIC WATER WITH ALGINIC ACID

Raw Alginic Acid

Table B.7

Operational conditions for duplicated experiments with raw alginic acid (DOC: 4 mg/L C)

Experiments	TMP (kPa)	Crossflow velocity (cm/s)	Clean water specific flux, J_{so} (L/m^2-hr-kPa)
#1	89.0 – 9 5.9	9.85	1.67
#2	91.0 – 97.4	10.4	1.65
#3	91.0 – 95.2	10.3	1.79

Table B.8

Water quality of duplicated experiments with raw alginic acid (DOC: 4 mg/L C)

Experiments	Samples	pH	Ca^{2+} (mg/L)	Mg^{2+} (mg/L)	DOC (mg/L)
#1	Feed	8.61	68.6	16.8	4.5
	Filtrate	8.58	65.3	16.8	0.4
#2	Feed	8.41	70.4	16.7	4.7
	Filtrate	8.48	66.1	16.9	0.3
#3	Feed	NA	102.0	16.4	4.1
	Filtrate		95.1	16.3	0.1

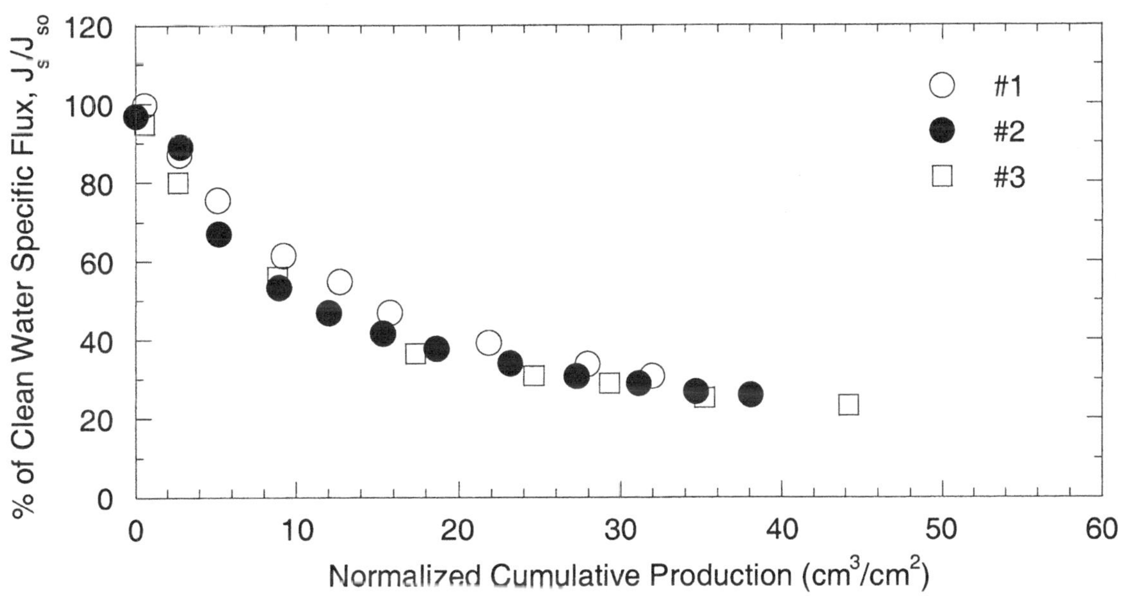

Figure B.4 Flux decline: duplicated experiments with raw alginic acid (4 mg/L C)

Softened Alginic Acid at 170 mg/L CaO

The flux decline showed a discrepancy between two sets of flux data at the lime dose of 170 mg/L with slightly different soda ash addition (as shown in Figure B.5). The data set with the lesser flux decline showed that the flux was better than the one at 125 mg/L of lime, whereas the data set with the higher flux decline was worse than that at 125 mg/L of lime. Several explanations could be proposed for causes of this discrepancy. The data with the worst flux had the higher clean water specific flux (i.e., approximately 1.5 times) compared to the other experiment. Although fluctuation in the clean water specific flux is accounted for in the normalization as the normalized cumulative production, perhaps the normalization is not enough to consider entirely the effects of differences in the clean water specific flux between membrane sheets. The data with the worse flux had higher concentrations of Ca^{2+} because of lower soda ash dose. In addition, DOC in the feed water was higher in the experiment with the worse flux. The data with the worse flux obtained only 18% DOC removal and that with better flux had 25% DOC removal in the softening pretreatment. Considering that a portion of alginic acid is critical to membrane fouling, the different DOC removal may be the main reason for the discrepancy.

Table B.9

Operational conditions for duplicated experiments with softened alginic acid at 170 mg/L CaO (DOC: 4 mg/L C)

Experiments	Soda ash (mg/L CaO)	TMP (kPa)	Crossflow velocity (cm/s)	Clean water specific flux, J_{so} (L/m^2-hr-kPa)
#1	45	84.8 – 93.8	9.92	1.76
#2	20	86.9 – 92.4	10.3	2.49

Table B.10

Water quality of duplicated experiments with softened alginic acid at 170 mg/L CaO (DOC: 4 mg/L C)

Exp.	Soda ash (mg/L CaO)	Sample	pH	Turbidity (NTU)	Ca^{2+} (mg/L)	Mg^{2+} (mg/L)	DOC (mg/L)
#1	45	Raw	8.42	NA	50.9	14.8	3.9
		Softened	11.06	14.5	16.5	5.1	NA
		Feed	10.88	4.25	12.6	3.9	2.9
		Filtrate	10.80	0.016	9.4	2.3	0.3
#2	20	Raw	8.42	NA	55.5	16.6	4.3
		Softened	11.06	14.5	25.4	4.7	3.8
		Feed	10.85	4.60	20.6	4.1	3.5
		Filtrate	10.80	0.018	16.1	1.7	D.L.

D.L.: lower than detection limit

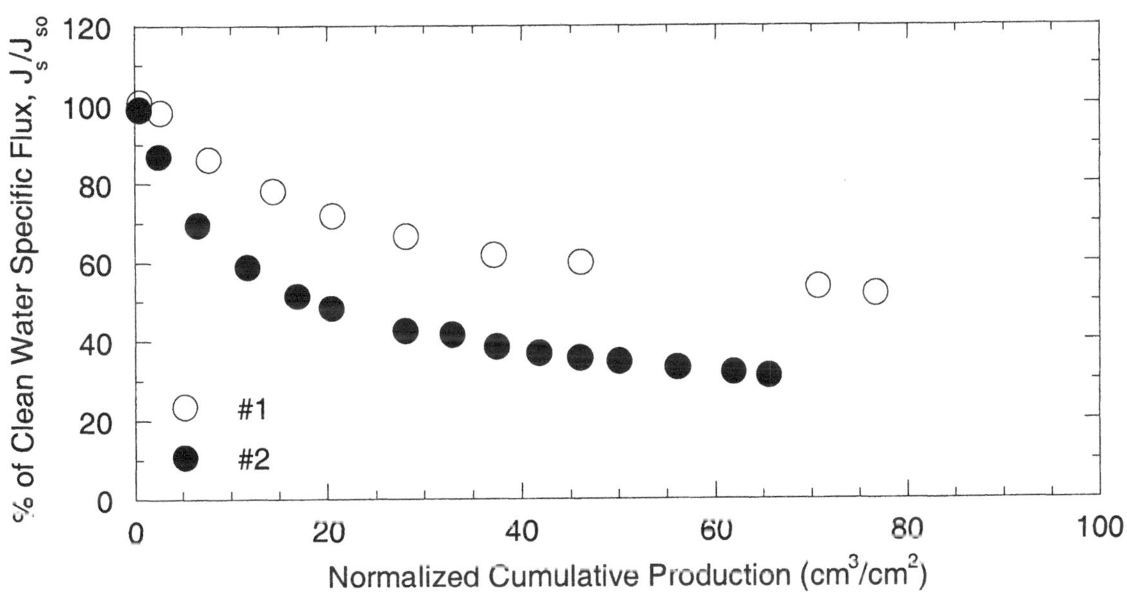

Figure B.5 Flux decline: duplicated experiments with softened alginic acid at 170 mg/L CaO (4 mg/L C)

B.4 SYNTHETIC WATER WITH CLAY AND RAW DEXTRAN

Table B.11
Operational conditions for duplicated experiments with kaolin and raw dextran (DOC: 4mg/L C, Turbidity: > 500 NTU)

Experiments	TMP (kPa)	Crossflow velocity (cm/s)	Clean water specific flux, J_{so} (L/m^2-hr-kPa)
#1	89.7-96.6	10.1	1.54
#2	95.9-100.7	10.0	1.20

Table B.12
Water quality of duplicated experiments with kaolin and raw dextran (DOC: 4 mg/L C, Turbidity: > 500 NTU)

Experiments	Sample	pH	Turbidity (NTU)	Ca^{2+} (mg/L)	Mg^{2+} (mg/L)	DOC (mg/L)
#1	Feed Filtrate	8.34 8.38	560.3 0.02	45.4 46.5	16.7 16.9	4.2 3.2
#2	Feed Filtrate	NA	NA	62.6 52.9	17.1 16.7	NA 2.7

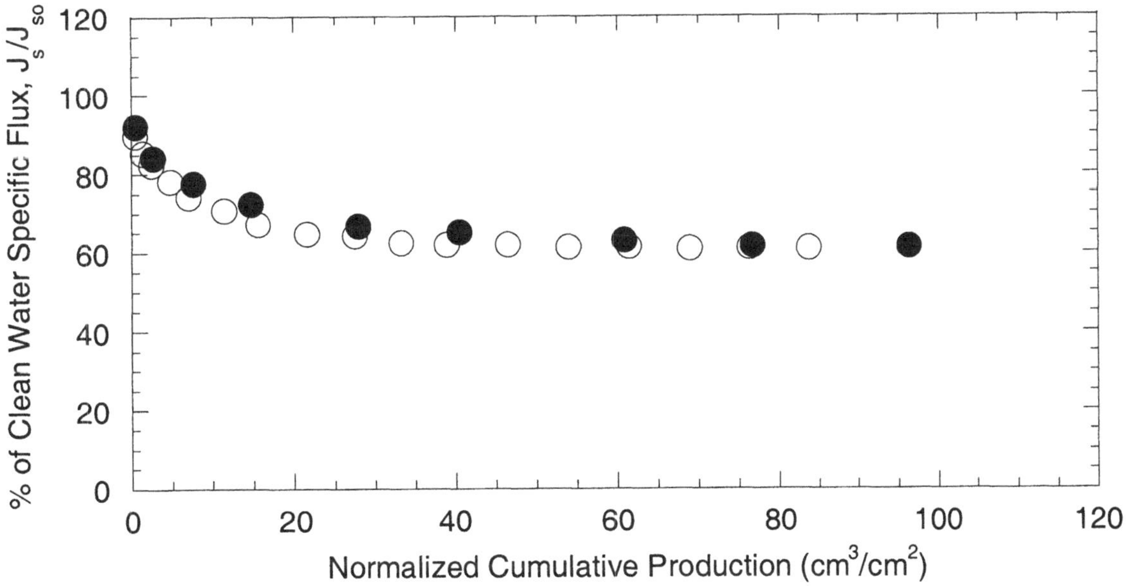

Figure B.6 Flux decline: duplicated experiments with kaolin and raw dextran (DOC: 4 mg/L C, Turbidity: >500 NTU)

B.5 NATURAL WATER: LAKE AUSTIN WATER

Softened Lake Austin water at 125 mg/L CaO

Table B.13
Operational conditions for duplicated experiments with softened Lake Austin water at 125 mg/L CaO

Experiments	TMP (kPa)	Crossflow velocity (cm/s)	Clean water specific flux, J_{so} (L/m^2-hr-kPa)
#1	92.4 – 105.5	9.3	2.07
#2	90.3 – 107.6	10.1	0.88

Table B.14
Water quality of duplicated experiments with softened Lake Austin water at 125 mg/L CaO

Exp.	Samples	pH	Turbidity (NTU)	Ca^{2+} (mg/L)	Mg^{2+} (mg/L)	DOC (mg/L)	UV_{254} (1/cm)	SUVA (L/m-mg)
#1	Raw	8.28	5.7	67.0	21.3	3.6	0.097	2.7
	Softened	10.91	28.7	-	-	-	-	-
	Feed	10.42	19.2	22.2	20.8	3.1	0.074	2.4
	Filtrate	10.56	0.04	13.6	19.8	2.9	0.054	1.8
#2	Raw	8.22	3.3	64.1	20.0	3.6	0.094	2.6
	Softened	10.70	27.5	20.4	19.9	2.6	0.081	-
	Feed	9.77	20.3	22.3	20.7	2.8	0.069	2.6
	Filtrate	9.74	0.51*	14.7	19.4	2.2	0.049	2.3

* Contamination from a sampling bottle

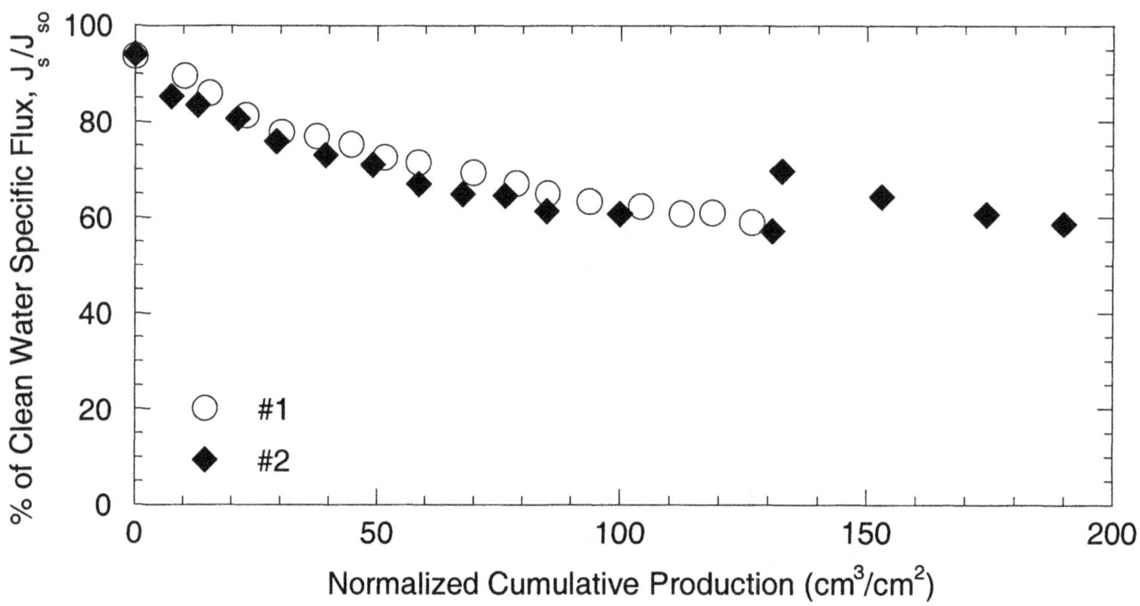

Figure B.7 Flux decline: duplicated experiments with softened Lake Austin water at 125 mg/L CaO (Note: the same graph was shown as Figure 4.3 in the main body of the text.)

APPENDIX C
EXPERIMENTS WITH REGENERATED CELLULOSE

C.1 EFFECTS OF RAW WATER CHARACTERISTICS ON FOULING

Experiments with regenerated cellulose membranes were performed with two natural waters. Transmembrane pressure (TMP) was maintained between 91.7 – 96.6 kPa for the experiment with Lake Austin water. The fluctuation of the TMP was relatively large for the experiment with Missouri River water (*i.e.*, 91 – 116.6 kPa) because of high solids concentration. The clean water specific flux was quite similar in both experiments, *i.e.*, 0.45 L/m^2-hr-kPa for Lake Austin water and 0.51 L/m^2-hr-kPa for Missouri River water. The flux decline from two water sources was significantly different, which was likely due to the differences in turbidity.

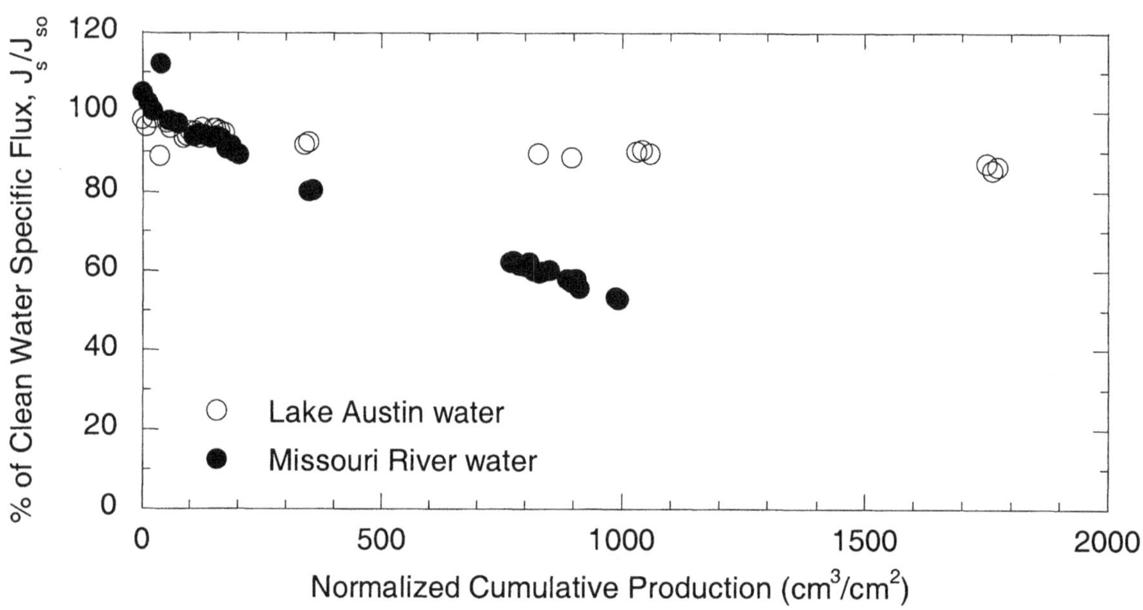

Figure C.1 Flux decline: effects of raw water characteristics

C.2 EFFECTS OF TRANSMEMBRANE PRESSURE ON FLUX DECLINE

Two transmembrane pressures (TMP) were investigated with regenerated cellulose membrane using Lake Austin water. The crossflow velocity was maintained at approximately 10 cm/s. As shown in Figure C.2, virtually no difference in flux decline was observed.

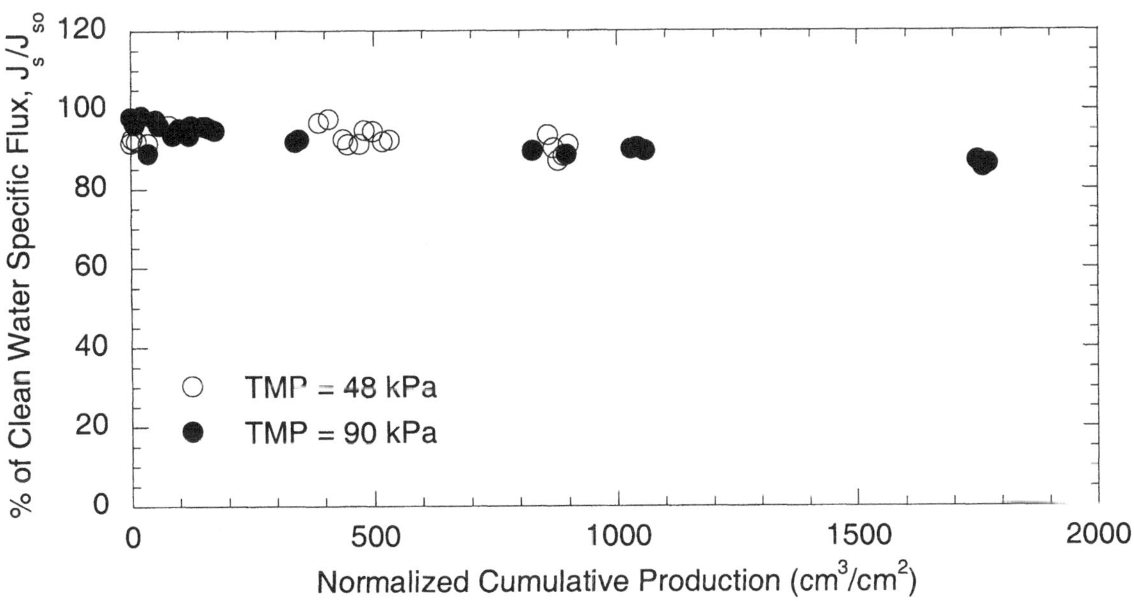

Figure C.2 Flux decline: effects of transmembrane pressure

C.3 EFFECTS OF SOFTENING PROCESSES ON FOULING

Regenerated cellulose membranes were used to investigate three softening configurations: raw water, softened water with settling, and flocculated water (without settling) followed by ultrafiltration. The operational conditions in these experiments were maintained well in the proper ranges, including TMP, crossflow velocity, and the clean water specific flux. The flux decline in Figure C.3 shows little change with different configurations.

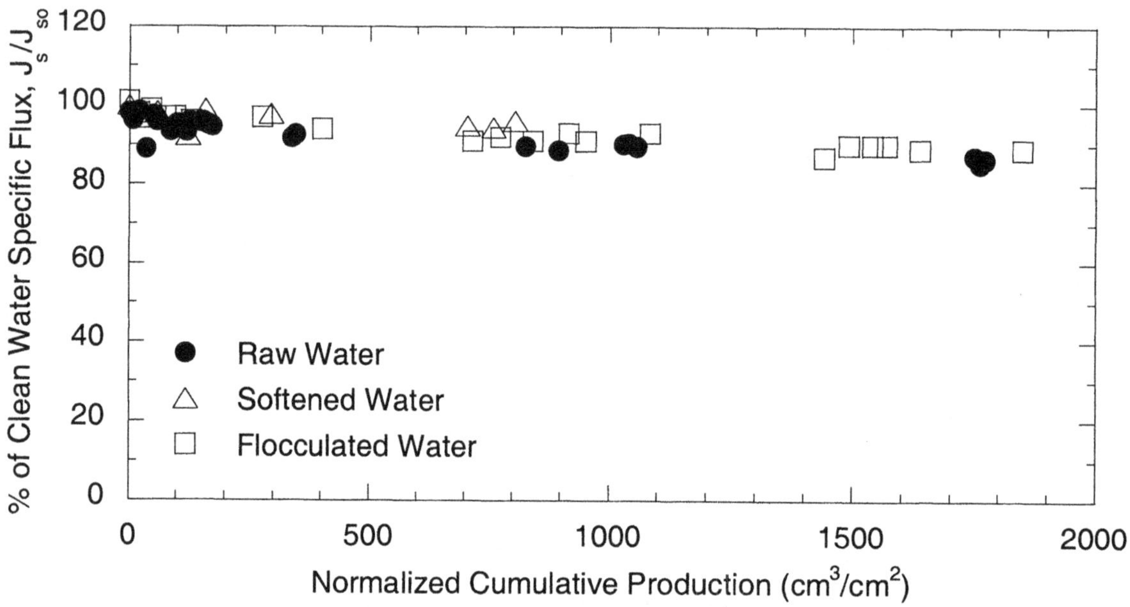

Figure C.3 Flux decline: effects of various softening configurations on fouling

Experiments with regenerated cellulose membranes showed little difference between experiments performed at different softening conditions, and also showed very little flux decline. The small flux decline using this membrane excluded the possibility of examining advantages of the softening process, which can remove particles and NOM thus can reduce membrane fouling. Therefore, no further investigation with regenerated cellulose membrane material was performed in this research.

REFERENCES

Adham, S.S., V.L. Snoeyink, M.M. Clark, and J. Bersillon. 1991. Predicting and Verifying Organics Removal by PAC in an Ultrafiltration System. Jour. AWWA, 83(12): 81-91.

Alexander, H.J. and M.A. McClanahan. 1975. Kinetics of Calcium Carbonate Precipitation in Lime-Soda Ash Softening. Jour. AWWA, 67(11): 618-621.

APHA, AWWA, and WEF (American Public Health Association, American Water Works Association, and Water Environment Federation). 1998. Standard Methods for the Examination of Water and Wastewater. 20th ed. Washington, D.C.: APHA.

ASTM (American Society for Testing and Materials). 1995. Annual Book of Standards, D4189-95. Philadelphia, Pa.: American Society for Testing and Materials.

AWWA Membrane Technology Research Committee. 1998. Committee Report: Membrane Processes. Jour. AWWA, 90(6): 91-105.

Bacchin, P, P. Aimar, and V. Sanchez. 1996. Influence of surface interaction on transfer during colloid ultrafiltration. Jour. Mem. Sci., 115 (1): 49-63.

Belfort, G., R.H. Davis, and A.L. Zydney. 1994. The behaviour of Suspensions and Macromolecular Solutions in Crossflow Microfiltration. Jour. Mem. Sci., 96: 1-58.

Bergman, R.A. 1996. Cost of Membrane Softening in Florida. Jour. AWWA, 88(5): 32-43.

Bouchard, C.R., J. Jolicoeur, and P. Kouasio. 1997. Study of Humic Acid Adsorption on Nanofiltration Membranes by Contact Angle Measurements. The Canadian Jour. of Chem. Eng., 75(April) 339:345.

Bowen, W.R. and F. Jenner. 1995. Theoretical Descriptions of Membrane Filtration of Colloids and Fine Particles: An Assessment and Review. Advances in Colloid and Interface Science, 56: 141-2000.

Braghetta, A., J.G. Jacangelo, S. Chellam, M.L. Hotaling, and B.A. Utne. 1997. DAF Pretreatment: Its Effects on MF Performance. Jour. AWWA, 89(10): 90-101.

Braghetta, A., M. Price, C. Hentz, and E. Stevenson. 2000. Impact of Clarification and Adsorption Pretreatment on UF Operational Performance and Taste and Odor Removal. In Proc. of the annual AWWA Conference. Denver, CO.: AWWA.

Briggs, D. 1983. Chapter 9. Applications of XPS in polymer Technology. In Practical Surface Analysis by Auger and X-ray Photoelectron Spectroscopy. Edited by D. Briggs and M.P. Seah. New York: NY., John Wiley & Sons, Ltd.

Briggs, D and J.C. Riviere. 1983. Chapter 3. Spectral Interpretation. In Practical Surface Analysis by Auger and X-ray Photoelectron Spectroscopy. Edited by D. Briggs and M.P. Seah. New York: NY., John Wiley & Sons, Ltd.

Bruchet, A., C. Rousseau, J. Mallevialle. 1990. Pyrolysis-GC-MS for Investigating High-Molecular-Weight THM precursors and Other Refactory Organics. Jour. AWWA., 82(9): 66-74.

Buffle, J. 1990. Complexation Reactions in Aquatic Systems: An Analytical approach. West Sussex, England: Ellis Horwood Limited.

Buffle, J., K.J. Wilkinson, S. Stoll, M. Filella, and J. Zhang. 1998. A generalized description of aquatic colloidal interactions: The three-colloidal component approach. Environ. Sci. Technol., 32(19): 2887-2899.

California Department of Health Services (2001). California Surface Water Treatment Alternative Filtration Technology Demonstration Report, ed. by R.H. Sakaji, available at http://www.dhs.ca.gov/ps/ddwem/publications/pubindex.htm

Campos, C., L. Schimmoller, B.J. Marinas, V.L. Snoeyink, I. Baudin, and J. Laine. 2000. Adding PAC to Remove DOC. Jour. AWWA, 92(8): 69-83.

Carroll, T., S. King, S.R. Gray, B.A.Bolto and N.A. Booker. 2000. The Fouling of Microfiltration Membranes by NOM After Coagulation Treatment. Wat. Res., 34(11): 2861-2868.

Champlin, T.L. 2000. Modeling Natural Organic Matter Adsorption and Particle Deposition by Spiral-Wound Nanofiltration Membrane Elements. In Proc. of the annual AWWA Conference. Denver, CO.: AWWA.

Chang, Y. 1996. A Combined Iron Oxide Adsorption and Ultrafiltration Process for Antural Organic Matter Removal and Fouling. Ph. D. diss., The University of Washington, Seattle, WA.

Chellam, S., J.G. Jacangelo, T.P. Bonacquisti, and B.A. Schauer. 1997. Effect of Pretreatment on Surface Water Nanofiltration. Jour. AWWA, 89(10): 77-89.

Chellam, S., C.A. Serra, and M.R. Wiesner. 1998. Estimating Costs for Integrated Membrane Systems. Jour. AWWA, 90(11): 96-104.

Cheryan, M. 1986. Ultrafiltration Handbook. Lancaster: Technomic Publishing Co. Inc.

Childress, A.E. and M. Elimelech. 1996. Effect of Solution Chemistry on the Surface Charge of Polymeric Reverse Osmosis and Nanofiltration Membranes. Jour. Mem. Sci., 119: 253-268.

Cho, J. 1998. Natural Organic Matter (NOM) Rejection by, and Flux-decline of, Nanofiltration (NF) and Ultrafiltration (UF) Membranes. Ph. D. diss., The University of Colorado, Boulder, CO.

Chow, C.W.K., van Leeuwen, J.A., Drikas, M., Fabris, R., Spark., K.M., Page, D.W. 1999. The impact of the character of natural organic matter in conventional treatment with alum. Water Science and Technology, 40(9), 97-104

Collins, M.R., G.L. Amy, and P.H. King. 1985. Removal of Organic Matter in Water Treatment. Jour. Environ. Eng., 111(6): 850-864.

Collins M.R., G.L. Amy, C. Steelink. 1986. Molecular-Weight Distribution, Carboxylic Acidity, and Humic Substances Content of Aquatic Organic-Matter - Implications for Removal during Water Treatment. Env. Sci. Tech., 20(10): 1028-1032.

Cleveland, C.T. 1999. Big Advantages in Membrane Filtration. Jour. AWWA, 91(6): 10.

Crozes, G.F., J.G. Jacangelo, C. Anselme, and J.M. Laine. 1997. Impact of Ultrafiltration Operating Conditions on Membrane Irreversible Fouling. Jour. Mem. Sci. 124: 63-76.

Folk, R. L. 1974. The natural history of crystalline calcium carbonate: Effect of magnesium content and salinity. Journal of Sedimentary Petrology., 44: 1: 40-53.

Fradin, B and R.W. Field. 1999. Crossflow Microfiltration of Magnesium Hydroxide Suspensions: Determination of Critical Fluxes, Measurement and Modeling of Fouling. Separation and Purification Technology, 16(1): 25-45.

Freeman, S., M. Horsely, A. Hess. 2000. Experiences with Submerged and Encased Microfiltration and Ultrafiltration Membranes for Surface Water Treatment. In Proc. of the Annual AWWA conference. Denver, Colo.: AWWA.

Handley, A. 1993. Chromatographic methods. In B. J. Hunt and M. I. James, Eds., Polymer Characterization. New York. Blackie Academic & Professional.

Harmant, P. and P. Aimar. 1998. Coagulation of colloids in a boundary layer during cross-flow filtration. Colloids and Surfaces A:., 138 (2-3): 217-230

Hong, S. K. and Elimelech, M. 1997. Chemical and Physical Aspects of Natural Organic Matter (NOM) fouling of nanofiltration. Jour. Mem. Sci., 132: 159-181.

Hwang, C.J., Sclimenti, M.J., Krasner, S.W., Croue, J., Bruchet, A., Amy, G.L. 1999. Characterization and reactivity of hydrophilic natural organic matter (NOM) in a low-humic water. In Proc. 1999 Water Quality Technology Conference, AWWA, held in Tampa, FL Oct. 31-Nov. 3, 1999.

Jacangelo, J.G., J. Laine, K.E. Carns, E.W. Cummings, and J. Mallevialle. 1991. Low-Pressure Membrane Filtration for Removing Giardia and Microbial Indicators. Jour. AWWA, 83(9): 97-106.

Jacangelo, J.G., J. Laine, E.W. Cummings, and S.S. Adham. 1995. UF with Pretreatment for Removing DBP Precursors. Jour. AWWA, 87(3): 100-112.

Jönsson, C., A, Jönsson. 1995. Influence of the Membrane Material on the Adsorptive Fouling of Ultrafiltration Membranes. Jour. Mem. Sci., 108: 79-87.

Jucker, C., and M.M. Clark. 1994. Adsorption of Aquatic Humic Substances on Hydrophobic Ultrafiltration Membranes. Jour. Mem. Sci., 97: 37-52.

Judkins, Jr., J.F. and J.S. Hornsby. 1978. Color Removal from Textile Dye Waste using Magnesium Carbonate. Jour. WPCF, 50(11): 2446-2456.

Karimi, A.A., J.C. Vickers, and R.F. Harasick. 1999. Microfiltration Goes Hollywood: the Los Angeles Experience. Jour. AWWA, 91(6): 90-103.

Kavanaugh M.C. 1978. Modified Coagulation for Improved Removal OF Trihalomethane Precursors. Jour. AWWA, 70 (11): 613-620.

Kelley, W.A. and R.A. Olson. 1999. Selecting MF to Satisfy Regulation. Jour. AWWA, 91(6): 52-63.

Kim, K.J. and A.G. Fane. 1994. Low Voltage Scanning Electron Microscopy in Membrane Research. Jour. Mem. Sci., 88: 103-114.

Kim, K.J., A.G. Fane, C.J.D. Fell, and D.C. Joy. 1992. Fouling Mechanisms of Membranes during Protein Ultrafiltration. Jour. Mem. Sci., 68: 79-91.

Kim, K.J., A.G. Fane, C.J.D. Fell, T. Suzuki, M.R. Dickson. 1990. Quantitative Microscopic Study of Surface Characteristics of Ultrafiltration Membranes. Jour. Mem. Sci., 54: 89-102.

Krasner, S.W., J. Croue, J. Buffle, and E.M. Perdue. 1996. Three Approaches for Characterizing NOM. Jour. AWWA, 88(6): 66-79.

Lahoussine-Turcaud, V., M.R. Wiesner, and J. Bottero. 1990. Fouling in Tangential-Flow Ultrafiltration: The Effect of Colloid Size and Coagulation Pretreatment. Jour. Mem. Sci., 52: 173-190.

Lahoussine Turcaud, V., M.R. Wiesner, J. Bottero, and J. Mallevialle. 1990. Coagulation in Pretreatment for Ultrafiltration of a Surface Water. Jour. AWWA, 82(12): 76-81.

Laine,J.M, M.M. Clark, and J. Mallevialle. 1990. Ultrafiltration of Lake Water: Effect of Pretreatment on the Partitioning of Organics, THMFP, and Flux. Jour. AWWA, 82(12): 82-87.

Laine, J.M, J.P. Hagstrom, M.M. Clark, and J. Mallevialle. 1989. Effect of Ultrafiltration Membrane Composition. Jour. AWWA, 91(11): 61-67.

Lawler, D.F. 1997. Precipitation or Solid/Solution Equilibrium, Course Notes, CE386.5L Physical-Chemical Processes in Water and Wastewater Treatment, Environmental and Water Resources Engineering, Department of Civil Engineering, The University if Texas at Austin.

Liao, M.Y., and S.J. Randtke. 1985. Removing Fulvic Acid by Lime Softening. Jour. AWWA, 77(10): 78-88.

Liao, M.Y., and S.J. Randtke. 1986. Predicting the Removal of Soluble Organic Contaminants by Lime Softening. Wat. Res., 20(1): 27-35.

Lin, C., T. Lin, O.J. Hao. 2000. Effects of Humic Substance Characteristics on UF Performance. Wat. Res., 34(4): 1097-1106.

Lloyd, D. R., (Professor of Chemical Engineering, U. of Texas), personal communication, October, 1999.

Lykins, Jr. B.W. and R. M. Clark. 1989. Trihalomethane Precursor and Total Organic Carbon Removal by Conventional Treatment and Carbon. In Aquatic Humic Substances: Influence on Fate and Treatment of Pollutants. Edited by I.H. Suffet and P. MacCarthy. Washington, DC.: American Chemical Society.

Mackey, D. E. 2000. Fouling of ultrafiltration and nanofiltration membranes by dissoloved organic matter. Doctor of Philosophy, Rice University.

Mackey, D. E. and M.R. Wiesner. 1999. Characterization of Irreversible Fouling of UF and NF Membranes by DOM. In Proc. of the AWWA Membrane Technology Conference. Long Beach, CA.: AWWA.

Mallevialle, J., C., Anselme, and O. Marsigny. 1989. Effects of Humic Substances on Membrane Processes. Aquatic Humic Substances. Proceedings of the 193rd meeting of the American Chemical Society, Denver, CO., 749-767.

Membrex company web page, http://www.membrex.com.

Moulder, J.F. , W.F. Stickle, P.E. Sobol, and K.D. Bomben, 1992. Handbook of X-ray Photoelectron Spectroscopy. Eden Prairie, Minnesota: Physical Electronics, Inc.

Munro, H. S. and S. Singh, 1993. The characterization of polymer surfaces by XPS and SIMS. In Polymer Characterization. Edited by B. J. Hunt and M.I. James. Glasgow, UK: Blackie Academic & Professional.

Nancollas, G.H. and M.M. Reddy, 1974. Crystal Growth Kinetics of Minerals Encountered in Water Treatment Process. In Aqueous- Environmental Chemistry of Metals. Edited by A.J. Rubin. Ann Arbor, Mich.: Ann Arbor Science.

Nilson, J.A. and F.A. DiGiano. 1996. Influence of NOM Composition on Nanofiltration. Jour. AWWA, 88(5): 53-66.

NIST (National Institute for Standards and Technology). 2002. NIST X-ray Photoelectron Spectroscopy Database Version 3.0. Gaithersburg, MD.

Oldani, M. and G. Schock. 1989. Characterization of Ultrafiltration Membranes by Infrared Spectroscopy, ESCA, and Contact Angle Measurements. Jour. Mem. Sci., 43: 243-258.

Oldham, W.K. and R.J. Rush. 1978. Color Removal in Kraft Mill Wastewaters with Magnesium. Jour. WPCF, 50(1): 40-45.

Owen, D.M., G.L. Amy, Z.K. Chowdhury, R. Paode, G.McCoy, and K. Viscosil. 1995. NOM Characterization and Treatibility. Jour. AWWA, 87(1): 46-63.

Palacio, L., P. Pradanos, A. Hernandez, M.J. Ariza, J. Benavente, M. Nystrom. 2001. Phase-contrast scanning force microscopy and chemical heterogeneity of GR polysulfone ultrafiltration membranes. Appl. Phys. A, 73: 555-560.

Patel, N.M., D.W. Dwight, J.L. Hedrick, D.C. Webster, and J.E. McGrath. 1988. Surface and bulk phase separation in block copolymers and their blends – polysulfone/polysiloxane. Macromolecules, 21(9): 2689-2696.

Peters, G.H., E.R. Baumann, and M.A. Larson. 1989. Effects of Various Parameters on the Thickening of Softening Plant Sludges. Jour. AWWA, 81(3): 74-84.

Pontie, M., X. Chasseray, D. Lemordant, and J.M. Laine. 1997. The Streaming Potential Method for the Characterization of Ultrafiltration Organic Membranes and the Control of Cleaning Treatments. Jour. Mem. Sci., 129:125-133.

Ralls, S.M. 1999. Enhanced softening for disinfection by-product precursor control. Master's Thesis, Environmental and Water Resources Engineering, Department of Civil Engineering, The University of Texas at Austin.

Randtke, S.J. 1988 Organic Contaminant Removal by Coagulation and Related Process Combinations. Jour. AWWA, 80(5): 40-56.

Randtke, S.J., R.C. Hoehn, W.R. Knocke, A.M. Dietrich, B.W. Long, and N. Wang. 1994. A Comprehensive Assessment of DBP Precursor Removal by Enhanced Coagulation/Softening. In Proc. of the Annual AWWA Conference. New York, NY.: AWWA.

Randtke, S.J., C.E. Thiel, M.Y. Liao, and C.N. Yamaya. 1982. Removing Soluble Organic Contaminants by Lime-softening. Jour. AWWA, 74(4):1 92-202.

Reddy, M.M. and K.K. Wang. 1980. Crystallization of Calcium Carbonate in the Presence of Metal Ions. I. Inhibition by Magnesium Ion at pH 8.8 and 25°C. Jour. Crystal Growth, 50: 470.

Semmens, M.J. and A.B. Staples. 1986. The Nature of Organics Removed During Treatment of Mississippi River Water. Jour. AWWA. 78(2):76-81.

Singer, P.C., G.W. Harrington, J. Thompson, M. White. 1995. Enhanced Coagulation and Enhanced Softening for the Removal of Disinfection By-Product Precursors; an Evaluation," Submitted to the Disinfectants/ Disinfection By-Products Technical Advisory Workgroup of the Water Utility Council of the American Water Works Association, pp. 17-19.

Smith, K.N. 2001. Enhanced softening: Effects of raw water characteristics and chemical addition. Master's Thesis, Environmental and Water Resources Engineering, Department of Civil Engineering, The University of Texas at Austin.

Smith, K.N., C. Gerwe, J. Kweon, G.E. Speitel. Jr., D.F. Lawler. 2002. Guidelines for Enhanced Softening: Raw Water Characteristics, NOM removal, and DBP formation. In Proc. of the Annual AWWA Conference. New Orleans, LA.: AWWA.

Snoeyink, V.L. and D. Jenkins. 1980. Water Chemistry. New York; Wiley.

Sparks D.L. 1995. Environmental Soil Chemistry. San Diego, CA; Academic Press.

Sun, Z. 1998. Enhanced softening: Preliminary modeling and experimental design. Departmental Report, Environmental and Water Resources Engineering, Department of Civil Engineering, The University of Texas at Austin.

Tambo, N. and T. Kamei. 1989. Evaluation of Extent of Humic-substance Removal by Coagulation. Aquatic Humic Substances. Proceedings of the 193rd meeting of the American Chemical Society, Denver, CO., 749-767.

Thompson, J.D., M.C. White, G.W. Harrington, and P.C. Singer. 1997. Enhanced Softening: Factors Influencing DBP Precursor Removal. Jour. AWWA, 89(6): 94-105.

Thurman, E.M., R.L. Malcolm. 1981. Preparative Isolation of Aquatic Humic Substances. ES & T. 15(4): 463-466.

Van Gelder, A.M., Z.K. Chowdhury, D.F. Lawler. 1999. Conscientious Particle Counting. Jour. AWWA. 91(12):64-76.

Vik, E.A. and B. Eikebrokk. 1989. Coagulation Process for Removal of Humic Substances from Drinking Water. In Aquatic Humic Substances: Influence on Fate and Treatment of Pollutants. Edited by I.H. Suffet and P. MacCarthy. Washington, DC.: American Chemical Society.

White M.C., J.D. Thompson, G. W. Harrington, P.C. Singer. 1997. Evaluating criteria for enhanced coagulation compliance. Jour. AWWA, 89(5): 64-77.

Wiesner, M.R. and P. Aptel. 1996. Mass Transport and Permeate Flux and Fouling in Presure-Driven Processes. In Water Treatment Membrane Processes. Edited by J. Mallevialle, P.E. Odendaal, and M.R. Wiesner. New York, NJ.: McGraw-Hill.

Wiesner, M.R. and S. Chellam. 1992. Mass Transport Considerations for Pressure-Driven Membrane Processes. Jour. AWWA, 84(1): 88-95.

Wiesner, M.R., M.M. Clark, and J. Mallevialle. 1989. Membrane Filtration of Cogulated Suspensions. Jour. Env. Eng., 115(1): 20-40.

Wiesner, M.R., J. Hackney, S. Sethi, J.G. Jacangelo, and J.M. Laine. 1994. Cost Estimates for Membrane Filtration and Conventional Treatment. Jour. AWWA, 86(12): 33-41.

Wilkes, D.R., S. Chellam, T.P. Bonacquisti, B.W. Long, and J.G. Jacangelo. 1996. Conventional treatment as a pretreatment for nanofiltration: Impact on DBP precursor removal, fouling, and facilities design. In Proc. of the annual AWWA Conference. Toronto, Ontario: AWWA.

Wilkinson, K. J., E. Balnois, G. G. Leppard, J. Buffle. 1999. Characteristic features of the major components of freshwater colloidal organic matter revealed by transmission electron and atomic force microscopy. Colloids and Surface A:, 155: 287-310.

Wilkinson, K.J., A. Joz-Roland, J. Buffle. 1997. Different roles of pedogenic fulvic acids and aquagenic biopolymers on colloid aggregation and stability in freshwaters. Limnol. Oceanogr., 42(8): 1714-1724.

Yoon, S., C. Lee, K. Kim, and A. G. Fane. 1998. Effect of Calcium ion on the Fouling of Nanofilter by Humic Acid in Drinking Water Production. Wat. Res., 32(7):2180-2186.

ABBREVIATIONS

AA	atomic adsorption
Adj.	adjustment
AMW	apparent molecular weight
AWWA	American Water Works Association
AwwaRF	Awwa Research Foundation
°C	degree Celsius
CA	cellulose acetate
CA.	California
cm	centimeter
cm/s	centimeter per second
cm2	square centimeter
cm3	cubic centimeter
Co.	company
COD	chemical oxygen demand
conc.	concentration
CT	total carbonate concentration in moles/L
CT	Product of disinfectant concentration and time
Da	Dalton
DBP	Disinfection By-Product
D/DBP	Disinfection/Disinfection By-Product
DOC	dissolved organic carbon
ed.	edition
eds.	editors
eq/L	equivalent per liter
FA	fulvic acid
FESEM	field emission scanning electron microscopy
FL	Florida
GAC	granular activated carbon
GC	gas chromatography
g/cm·s	gram per centimeter per second
GWFA	ground water fulvic acid
HA	humic acid
IESWTR	Interim Enhanced Surface Water Treatment Rule
Jour.	Journal

kDa	kilodaltons
kPa	kilopascals
kV	kilovoltage
L	liter
LA	Lake Austin
L/m·hr	liter per meter per hour
L/m·hr·kPa	liter per meter per hour per kilopascal
MA	Massachusetts
MF	microfiltration
mg/L	milligram per liter
m2/g	square meter per gram
mM	milimole
MS	mass spectroscopy
MWCO	molecular weight cut-off
MWn	number-average molecular weight
MWw	weight-average molecular weight
μm	micrometer
NA	not available
NC	North Carolina
NCP	normalized cumulative production
NF	nanofiltration
NOM	natural organic matter
nm	nanometer
NTU	naphelometric turbidity unit
NV	Nevada
PA	polyamide
PAC	powdered activated carbon
PAC	Project Advisory Committee
PAN	polyacrylonitrile
PD	polydispersity
PE	polyethylene
PES	polyethersulfone
PHA	polyhydroxyaromatics
PP	polypropylene
Proc.	Proceedings
PS	polysulfone
PTFE	polytetrafluoroethylene
PVDF	polyvinylidene flouride
RO	reverse osmosis
SEC	size-exclusion chromatography
SEM	scanning electron microscopy

SOM	soil organic matter
SR	Suwannee River
St.	Saint
SUVA	specific ultra-violet absorbance
SWTR	Surface Water Treatment Rule
THM	trihalomethane
TMP	transmembrane pressure
TOC	total organic carbon
TX	Texas
UF	ultrafiltration
U.K.	United Kingdom
UKGSAHA	U.K. Geological Survey Aquatic Humic Acid
U.S.	United States
U.S.A.	United States of America
UV	ultra-violet
WA	Washington
XPS	X-ray photoelectron spectroscopy

IWA Publishing's authorised EU representative for General Product Safety Regulations is Diane D'Arras, 15 rue Duret, 75116 Paris, France, e-mail: safety@iwap.co.uk.

Printed and bound by CPI Group (UK) Ltd, Croydon, CR0 4YY
17/09/2026
02218793-0002